Lecture Notes of the Institute for Computer Sciences, Social Informatics and Telecommunications Engineering 550

The LNICST series publishes ICST's conferences, symposia and workshops.
LNICST reports state-of-the-art results in areas related to the scope of the Institute.
The type of material published includes

- Proceedings (published in time for the respective event)
- Other edited monographs (such as project reports or invited volumes)

LNICST topics span the following areas:

- General Computer Science
- E-Economy
- E-Medicine
- Knowledge Management
- Multimedia
- Operations, Management and Policy
- Social Informatics
- Systems

Lin Yun · Jiang Han · Yu Han

Editors

Advanced Hybrid Information Processing

7th EAI International Conference, ADHIP 2023
Harbin, China, September 22–24, 2023
Proceedings, Part IV

Editors
Lin Yun
Harbin Engineering University
Harbin, China

Jiang Han
Harbin Engineering University
Harbin, China

Yu Han
Harbin Engineering University
Harbin, China

ISSN 1867-8211   ISSN 1867-822X (electronic)
Lecture Notes of the Institute for Computer Sciences, Social Informatics
and Telecommunications Engineering
ISBN 978-3-031-50551-5   ISBN 978-3-031-50552-2 (eBook)
https://doi.org/10.1007/978-3-031-50552-2

This Springer imprint is published by the registered company Springer Nature Switzerland AG
The registered company address is: Gewerbestrasse 11, 6330 Cham, Switzerland

Paper in this product is recyclable.

Preface

We are delighted to introduce the proceedings of the 7th edition of the European Alliance for Innovation (EAI) International Conference on Advanced Hybrid Information Processing (ADHIP 2023). This conference brought together researchers, developers and practitioners around the world who are leveraging and developing advanced information processing technology. This conference aimed to provide an opportunity for researchers to publish their important theoretical and technological studies of advanced methods in social hybrid data processing, and their novel applications within this domain.

The technical program of ADHIP 2023 consisted of 108 full papers. The topics of the conference were novel technology for social information processing and real applications to social data. Aside from the high-quality technical paper presentations, the technical program also featured three keynote speeches. The three keynote speakers were Cesar Briso from Technical University of Madrid, Spain, Yong Wang from Harbin Institute of Technology, China, and Yun Lin from Harbin Engineering University, China.

Coordination with the steering chairs, Imrich Chlamtac, Shuai Liu and Yun Lin was essential for the success of the conference. We sincerely appreciate their constant support and guidance. It was also a great pleasure to work with such an excellent organizing committee team for their hard work in organizing and supporting the conference. In particular, the Technical Program Committee, led by our TPC Co-Chairs, Yun Lin, Ruizhi Liu and Shan Gao completed the peer-review process of technical papers and made a high-quality technical program. We are also grateful to the Conference Manager, Ivana Bujdakova, for her support and to all the authors who submitted their papers to the ADHIP 2023 conference.

We strongly believe that the ADHIP conference provides a good forum for all researchers, developers and practitioners to discuss all technology and application aspects that are relevant to information processing technology. We also expect that the future ADHIP conferences will be as successful and stimulating as indicated by the contributions presented in this volume.

Yun Lin

Organization

Organizing Committee

General Chair

Yun Lin — Harbin Engineering University, China

General Co-chairs

Zheng Dou — Harbin Engineering University, China
Yan Zhang — University of Oslo, Norway
Shui Yu — University of Technology Sydney, Australia
Joey Tianyi Zhou — Institute of High-Performance Computing, A*STAR, Singapore
Hikmet Sari — Nanjing University of Posts and Telecommunications, China
Bin Lin — Dalian Maritime University, China

TPC Chair and Co-chairs

Yun Lin — Harbin Engineering University, China
Guangjie Han — Hohai University, China
Ruolin Zhou — University of Massachusetts Dartmouth, USA
Chao Li — RIKEN-AIP, Japan
Guan Gui — Nanjing University of Posts and Telecommunications, China
Ruizhi Liu — Harbin Engineering University, China

Sponsorship and Exhibit Chairs

Yiming Yan — Harbin Engineering University, China
Ali Kashif — Manchester Metropolitan University, UK
Liang Zhao — Shenyang Aerospace University, China

Local Chairs

Jiang Hang	Harbin Engineering University, China
Yu Han	Harbin Engineering University, China
Haoran Zha	Harbin Engineering University, China

Workshops Chairs

Nan Su	Harbin Engineering University, China
Peihan Qi	Xidian University, China
Jianhua Tang	Nanyang Technological University, Singapore
Congan Xu	Naval Aviation University, China
Shan Gao	Harbin Engineering University, China

Publicity and Social Media Chairs

Jiangzhi Fu	Harbin Engineering University, China
Lei Chen	Georgia Southern University, USA
Zhenyu Na	Dalian Maritime University, China

Publications Chairs

Weina Fu	Hunan Normal University, China
Sicheng Zhang	Harbin Engineering University, China
Wenjia Li	New York Institute of Technology, USA

Web Chairs

Yiming Yan	Harbin Engineering University, China
Zheng Ma	University of Southern Denmark, Denmark
Jian Wang	Fudan University, China

Posters and PhD Track Chairs

Lingchao Li	Shanghai Dianji University, China
Jibo Shi	Harbin Engineering University, China
Yulong Ying	Shanghai University of Electric Power, China

Panels Chairs

Danda Rawat   Howard University, USA
Yuan Liu   Tongji University, China
Yan Sun   Harbin Engineering University, China

Demos Chairs

Ao Li   Harbin University of Science and Technology,
                       China
Guyue Li   Southeast University, China
Changbo Hou   Harbin Engineering University, China

Tutorials Chairs

Yu Wang   Nanjing University of Posts and
                       Telecommunications, China
Yi Zhao   Tsinghua University, China
Qi Lin   Harbin Engineering University, China

Technical Program Committee

Zheng Dou   Harbin Engineering University, China
Yan Zhang   University of Oslo, Norway
Shui Yu   University of Technology Sydney, Australia
Joey Tianyi Zhou   A*STAR, Singapore
Hikmet Sari   Nanjing University of Posts and
                       Telecommunications, China
Bin Lin   Dalian Maritime University, China
Yun Lin   Harbin Engineering University, China
Guangjie Han   Hohai University, China
Ruolin Zhou   University of Massachusetts Dartmouth, USA
Chao Li   RIKEN-AIP, Japan
Guan Gui   Nanjing University of Posts and
                       Telecommunications, China
Zheng Ma   University of Southern Denmark, Denmark
Jian Wang   Fudan University, China
Lei Chen   Georgia Southern University, USA
Zhenyu Na   Dalian Maritime University, China
Peihan Qi   Xidian University, China
Jianhua Tang   Nanyang Technological University, Singapore

Congan Xu	Naval Aviation University, China
Ali Kashif	Manchester Metropolitan University, UK
Liang Zhao	Shenyang Aerospace University, China
Weina Fu	Hunan Normal University, China
Danda Rawat	Howard University, USA
Yuan Liu	Tongji University, China
Yi Zhao	Tsinghua University, China
Ao Li	Harbin University of Science and Technology, China
Guyue Li	Southeast University, China
Lingchao Li	Shanghai Dianji University, China
Yulong Ying	Shanghai University of Electric Power, China

Contents – Part IV

Information Theory and Coding for Social Information Processing, Civilian Industry Technology Tracks

Intelligent Monitoring Method for Static Comfort of Ejection Seat Based on Human Factors Engineering

Jinshou Shi[✉] and Chao Ma

College of Aeronautical Engineering, Beijing Polytechnic, Beijing 100176, China
shijinshou@bpi.edu.cn

Abstract. In order to determine the factors that affect the static comfort of ejection seats, an intelligent monitoring method for static safety of ejection seats based on human factors engineering is proposed. Human factors engineering is used to analyze human motion, and different percentiles of human body size are used to simulate the leg motion range, so as to determine the travel and adjustment amount of the pedal.Combined with the size of the human body, the position of the seat reference point and the comfort Angle of each joint of the human leg, the comfort range of the pedal shaft is solved, and the pedal is optimized. The finite element model of the ejection seat is constructed by human factors engineering, and the linear elastic constitutive relationship is used as the material constitutive relationship of the ejection seat. The material density, Young's modulus, Poisson's ratio, shear modulus, bulk modulus and other parameters are defined to analyze the impact load of the ejection seat when it is ejected. The results indicate that the finite element model based on ergonomics is suitable for static comfort monitoring of ejection seats. The increase in seat cushion thickness has a certain trend of reducing the maximum and average pressure in certain areas of the seat cushion, and has almost no effect on the contact area. Increasing the protrusion of the waist can increase the maximum pressure on the waist, but it will reduce the contact area at the shoulder position. In the development of comfort, attention should be paid to the matching of the waist and shoulders to ensure the overall comfort of the backrest.

Keywords: Human Factors Engineering · Human Movement · Ejection Seat · Static Comfort · Finite Element Model

1 Introduction

Military aircraft has always played a very important role in the development history of air navigation. The two world wars showed that aircraft played a very important role in military affairs, but it was also very likely to be destroyed. In order to save aircrew after aircraft damage, special equipment must be used to enable them to get off the aircraft smoothly and land safely [1]. Long term flight missions often performed by pilots will

L. Yun et al. (Eds.): ADHIP 2023, LNICST 550, pp. 3–19, 2024.
https://doi.org/10.1007/978-3-031-50552-2_1

lead to deep vein thrombosis, which will lead to pressure ulcers, delayed response, and reduced operating ability, which will seriously affect the health of pilots and reduce the effectiveness of flight operations. Safety issues will be given priority in the design of ejection seats, so the comfort improvement of ejection seats is very limited, and the cushion is the most important part that can be improved and optimized, and can improve the comfort of ejection seats.

The uniform pressure distribution on the seat surface can reduce the probability of skin ulceration and tenderness. The local high pressure on the person seat interface will lead to the deformation of human soft tissue, which will hinder the blood circulation and nutrition supply, and make the human body feel uncomfortable or tired. Therefore, according to the physiological response, the stress distribution that is most suitable for sitting position can make the weight evenly distributed on the larger support surface, but not evenly distributed; And the body pressure distribution process should be smooth transition from small to large to avoid sudden changes. The maximum pressure at the ischial tubercle occurs on the seat surface, and the pressure gradually decreases outward from the ischial tubercle until the pressure is the minimum under the thigh that contacts the front edge of the seat surface. Because the sitting stress of the seat cushion affects the comfort performance of the seat to a large extent, the comfort performance of the seat can be significantly improved by improving the distribution of the sitting stress, so it is of great significance to study the sitting stress of the person seat interface for improving the comfort of the seat.

In domestic research, Han Yuhong et al. [2]. by designing the experiment, the independent variable is two different seats, and the dependent variable is the skin electricity reflex, heart rate variability, pupil diameter, and the user's subjective evaluation of the seat comfort. SPSS statistical analysis software is used to do independent sample t-test on the experimental data, and analyze their correlation. The results showed that there were significant differences in the changes of pupil diameter between the subjects sitting in two different seats, and there was no significant difference in the detection results of EDA and HRV.In the research of seat comfort evaluation based on psychophysiological measurement method, pupil index is relatively sensitive and has strong correlation with subjective evaluation. Under certain conditions, pupil data can be used to optimize seat comfort evaluation; There is no sufficient evidence to support that there is a significant correlation between the skin electric reflex and heart rate variability and seat comfort. In addition, the application of psychophysiological measurement methods such as eye tracking can, to a certain extent, reverse guide the improvement and design of seats. Longjiang et al. [3] can not establish an accurate seat comfort prediction model because the traditional BP neural network is sensitive to the initial value and easy to fall into the local optimal solution. In order to solve this problem, a method to predict the comfort of seats by using the BP neural network optimized by the artificial bee colony algorithm is proposed. 176 groups of pressure distribution sample data were obtained through the body pressure test, 89% of which were used as the training part of the model, and 11% of which were used as the model validation. Comparing the prediction results with the real values, the mean square error MSE of the ABC-BP prediction model was 0.0019, and the certainty coefficient $R \wedge 2$ was 0.946, which was 84.68% lower than the MSE obtained by the traditional BP neural network prediction model, and $R \wedge 2$ was 42.5%

higher. The results show that the prediction model of vehicle seat comfort established by BP neural network optimized by artificial bee colony algorithm is more stable and more accurate.

In foreign studies, Li M et al. [4]. By measuring the pressure distribution and EMG signal during short and long driving respectively, the driver's pressure index, EMG signal characteristic parameters and corresponding subjective evaluation are obtained. The changes of subjective and objective parameters of comfort, the differences in comfort and the body parts prone to fatigue were tested. The index screening was completed through correlation analysis. A new comprehensive weighting method, AHP restricted entropy method, is proposed to establish the mapping relationship between subjective comfort and objective indicators based on pressure distribution and physiological information during short-term and long-term driving. A quantitative evaluation method of driving comfort is obtained by applying pressure distribution, physiological information and subjective evaluation, which provides a theoretical basis for evaluating driving comfort.

In view of the lack of research on the comfort of ejection seat cushion, this paper applies ergonomics to the intelligent monitoring of static comfort of ejection seat based on the research on the stress distribution of vehicle seat, so as to improve the static comfort of ejection seat. In the research process of this paper, firstly, based on the three-dimensional coordinate system, the relative coordinate system fixed on each moving limb is established, and the human factors engineering is used to analyze the human motion and optimize the layout of the cockpit heading foot pedal. Then, considering the complex actual structure and use environment of the ejection seat, and trying to accurately calculate the impact dynamic characteristics of the seat, the linear elastic constitutive relationship is used as the material constitutive relationship of the ejection seat. Finally, by setting the boundary conditions, the finite element model of the ejection seat was constructed by human factors engineering, and the intelligent monitoring of the static comfort of the ejection seat was realized.

2 Design of Intelligent Monitoring Method for Static Comfort of Ejection Seat

2.1 Optimizing the Arrangement of Cockpit Heading Pedals

When the manikin simulates the pilot driving the aircraft, the pilot leans back against the seat, looks ahead, holds the steering stick with his right hand and the accelerator lever with his left hand. When simulating human movement, according to the state of the pilot when he is in a natural upright sitting position with head up view [5], the whole trunk is set to remain motionless, and only the hand and foot joints need to move within a reasonable range.

In human motion analysis, the position of the hand or foot in any space is determined by formula (1) by establishing a three-dimensional coordinate system (X, Y, Z) And then establish the relative coordinate system fixed on each moving limb (X_n, Y_n, Z_n) When the human body moves, the spatial position of the hand or foot in the base coordinate system is shown as follows:

$$\left\{ \begin{array}{c} X_n \\ Y_n \\ Z_n \end{array} \right\} = T_n \left\{ \begin{array}{c} X \\ Y \\ Z \end{array} \right\} + P_n \tag{1}$$

Where, P_n Is the coordinate system (X_n, Y_n, Z_n) Origin in base coordinate system (X, Y, Z) Coordinates in, T_n It is the transformation matrix from relative coordinate system to base coordinate system.

According to the analysis of human motion, the cockpit directional pedals are optimally arranged. For the pedal stroke and adjustment amount, the leg movement range can be determined only by simulating the human body size at different percentiles [6]. Combining the size of the human body, the position of the seat reference point and the comfort angle of each joint of the human leg, we can use Formula (2) to solve the comfort range of the pedal shaft and optimize the layout of the pedal.

$$\left. \begin{array}{l} \alpha + \beta + \gamma = 180° \\ B \times \sin \gamma = A \times \sin_\alpha + C + E \\ F = D + A \times \cos_\alpha + B \times \cos \end{array} \right\} \tag{2}$$

Among them, α Represents the thigh chamfer, β Represents the angle between the axis of the thigh and the axis of the lower leg, γ Represents the angle between the leg axis and the horizontal line passing through the pedal axis point, A Represents the thigh length, B Indicates the length of the lower leg, C Represents the longitudinal distance from the hip joint to the reference point, D Represents the transverse distance, E Represents the vertical distance from the reference point to the pedal, F The horizontal distance from the reference point to the pedal.

The finite element model of the ejection seat is built after the optimal layout of the cockpit directional pedals is completed.

2.2 Build Finite Element Model of Ejection Seat

2.2.1 Simulation Model

Human factors engineering takes the best matching of human machine environment system as an important goal pursued by the discipline [7], so that people under different conditions can work and live in an efficient, healthy, safe and comfortable manner, ensure high-quality operation and high work efficiency at the same time, and reduce or eliminate safety accidents caused by errors or misreading through comfort design.

Considering the complex actual structure and use environment of the ejection seat, and striving to accurately calculate the seat impact dynamic characteristics [8], the finite element model of the ejection seat is constructed using human factors engineering. The following basic assumptions are made in modeling:

- Except for the impact load bearing points of the structure, the entire structure only produces elastic deformation during the impact of the seat;
- The whole structure meets the continuity assumption and uniformity assumption;

- The materials used in the whole structure meet the basic assumption of isotropy.

In view of the complexity of the seat entity model, in order to effectively carry out the finite element modeling, the seat structure should be simplified and the main load-bearing structure should be retained. In the process of simplification, try to ensure that the overall stiffness is unchanged and the shape of key parts is consistent, so that the structure of key parts can be guaranteed more reliably. The simplified model is shown in Fig. 1.

Fig. 1. Finite Element Model of Ejection Seat

In the process of simulation calculation, the whole seat structure is properly simplified, and the main load-bearing structure in the actual impact process of the seat is retained, which greatly reduces the number of grids, shortens the calculation period and ensures the calculation accuracy. It can provide a reference for further analysis of ejection seats.

2.2.2 Material Parameters of Ejection Seat

The ejection seat is mainly made of aluminum alloy, which has a high strength to weight ratio. Based on the assumptions proposed in this paper and the study in this paper only considers the response of the seat in the linear elastic range [9], so the material constitutive relationship used in the calculation is linear elastic constitutive relationship.

In the linear elastic material mode, the material density needs to be defined ρ Young's modulus E Poisson's ratio μ, shear modulus G, bulk modulus K And other parameters. Only two of Young's modulus, Poisson's ratio, shear modulus and bulk modulus need to be defined, because the four have the following relationship:

$$G = \frac{E}{2(1 + \rho)} \tag{3}$$

$$K = \frac{E}{3(1 - 2\mu)} \tag{4}$$

There are two kinds of impact loads on the ejection seat. One is the acceleration impact load on the entire seat structure during the ejection process, and the other is the impact force on the seat basin during the pilot's ejection process. This paper mainly studies the dynamic response of the seat structure under these two impact loads.

It can be seen from the above brief introduction to the ejection seat that the seat basin and hook are the main parts of the seat when the machine falls. The two parts of the impact load are respectively loaded equivalently. The acceleration impact load borne by the seat during the actual crash acts on the entire seat structure in a vertical upward direction. In the simulation process, it is equivalent to the restraint hook part, and the entire acceleration load acts vertically downward on the seat structure; The impact load of passengers is applied to the seat basin. The impact load of passengers is equivalent to that applied to the upper surface of the front beam and the lower beam in a vertical downward direction. In the impact simulation process, due to the role of the passengers in the seat impact process, the dummy model with a mass of $m = 90$ kg is used to replace the passengers, so the force acting on the upper surface of the seat front beam and lower beam $F = ma$.

The displacement boundary condition is: since the slide rail is connected to the aircraft cabin through three pairs of pulleys, it is necessary to constrain the slide rail [10]. When constraining the front and rear directions and the left and right directions, the hook connected to the power ejection device is hung on the aircraft cabin. Therefore, full restraint is required on the arc surface of the hook, and the displacement in the three directions is zero.

According to the above process, the finite element model of the ejection seat is built to monitor its static comfort.

3 Experimental Analysis

3.1 Finite Element Model Verification

3.1.1 Comparison Between Simulation and Test Based on H-point Position

The H-point device is an important tool for human-machine detection. The H-point position value can correctly evaluate the relevant compression state of the seat and the measurement of the design parameters of the seat, so it can be used as an important reference index for the comparison of simulation and test results. In this paper, the H-point simulation test is introduced and compared.

Since the left and right sides of the dummy have been ensured to be symmetrical before the H-point test, so that the Y-coordinate of the H-point is kept at the design value of −355, it is only necessary to refer to the X and Z coordinate values. Table 1 summarizes the X and Z coordinate values of the seat design H-point position, the design values of the backrest angle parameters, and the simulation and test results.

As the ejection seat product is inferior to other seats, the design contour of the seat surface often has a certain degree of deviation from the real object, and the general requirements of the OEMs for the measured H-point of the ejection seat are within 10mm relative to the design R-point, so from the summary results in Table 1 above, it can be seen that the seat H-point results of this test meet the corresponding design

Table 1. Comparison of H-point position simulation and test

Test items	X	Z	Backrest angle
Design	1280	415	25°
Test	1281.8	413.6	25.3°
Simulation	1285.9	411.4	25.9°
Error	4.1 mm	−2.2 mm	0.6°

requirements, and the seat H-point position simulation and test errors. The position state errors of simulation and design are within a reasonable range. This conclusion shows that the comfort simulation model of ejection seat established in this paper can accurately simulate the H-point position of ejection seat and the relevant design parameters of the seat.

3.1.2 Comparison Between Simulation and Test of Body Pressure Distribution Based on HPM Device

As the main objective evaluation of comfort monitoring, the body pressure distribution is further compared with the simulation and test of the seat comfort simulation model established in this paper. This paper mainly introduces the process of comparing the simulation and test of seat body pressure distribution using HPM device.

(1) Simulation Analysis of HPM Device Seat Interface Pressure Results.

The simulation analysis steps of body pressure distribution based on HPM device are executed according to the relevant dummy placement method in the process of seat H-point position measurement. After the final calculation results are completed, the cloud diagram of body pressure results can be output in the software, as shown in Fig. 2.

(2) HPM device seat interface pressure result test.

Test preparation: HPM dummy device, body pressure test pad, seat tooling, steel ruler (front and rear position measurement of slide rail), seat sample.

Test steps: first install the test equipment and place it according to the placement method of the body pressure test pad to ensure that the body pressure test pad is stable and can read the pressure image normally; Then place the HPM_0th dummy device, complete the placement of HPM dummy according to the dummy placement method when measuring at H-point, and ensure that the HPM_The 0th dummy is placed in the middle of the seat, and the left and right positions of the dummy are measured by the three coordinate measuring device to make the height of the dummy consistent; Finally, observe the body pressure test results, and record the measured values after the pressure data is stable.

(a) Simulation Results of Seat Cushion Body Pressure Distribution

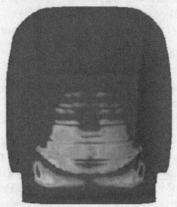

(b) Simulation results of backrest pressure distribution

Fig. 2. HPM based_Simulation results of 0th device body pressure distribution

The body pressure test results after final stabilization are shown in Fig. 3. Figure 3(a) shows the test results of seat cushion pressure, and Fig. 3(b) shows the test results of backrest pressure. Part of the profile features of the seat can be clearly identified through this cloud picture.

(3) Comparison of HPM device seat section pressure based on simulation and test.

First, the pressure distribution forms of simulation and test are analyzed. It can be seen that the maximum pressure in simulation and test both appears in the dummy's hip area. In the back pressure distribution cloud chart, the local maximum pressure appears on both sides of the lower part of the back. The back pressure distribution appears faults near the style line of the seat cover. Neither simulation nor test has contact above the shoulder, The pressure contact contour nephogram form of seat cushion and backrest is consistent.

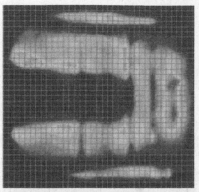

(a) Seat cushion body pressure distribution test results

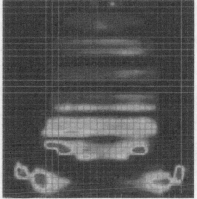

(b) Backrest pressure distribution test results

Fig. 3. Test Results of Body Pressure Distribution Based on HPM Device

Secondly, the transverse and longitudinal pressure distribution curves are compared and analyzed. The statistical method of the curve is shown in Fig. 4 below. The transverse pressure distribution curve takes the left and right transverse position data of the seat as the abscissa, and the collected pressure sum of each column as the ordinate; The longitudinal pressure distribution curve takes the longitudinal position data before and after the seat as the abscissa, and the pressure sum of each row as the ordinate. The pressure distribution of the seat can be more intuitively analyzed through the statistics of the transverse and longitudinal pressure distribution curves, the trend of the pressure gradient change can be counted, and the abnormal pressure distribution can be found.

Figure 5 shows the pressure distribution trend of the seat cushion and backrest in different directions under the statistical simulation and test according to the transverse and longitudinal pressure distribution curve of the seat cushion and backrest.

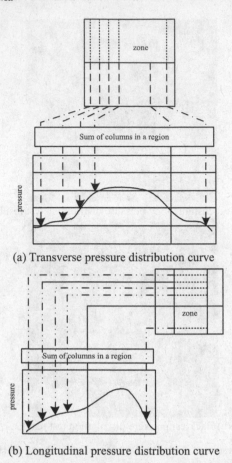

(a) Transverse pressure distribution curve

(b) Longitudinal pressure distribution curve

Fig. 4. Schematic Diagram of Horizontal and Longitudinal Pressure Distribution Curve of Body Pressure

From the comparison of the transverse and longitudinal pressure distribution curves, it can be seen that the curve distribution trends of simulation and test are consistent, and the pressure values are close. The longitudinal pressure distribution curve of the backrest fluctuates greatly, which is caused by the shape curve of the seat backrest, but the overall trend of simulation and test is still consistent.

Finally, in order to further and more intuitively compare and analyze the body pressure distribution results of simulation and test, the body pressure index values of the seat cushion and backrest in the statistical simulation and test, including the maximum contact pressure, average pressure, and contact area index values, are compared and summarized as shown in Table 2.

The results in the table show that the two are very close, and the maximum relative error is less than 10%. Therefore, the seat simulation model established in this paper can accurately reflect the H-point device seat body pressure distribution characteristics.

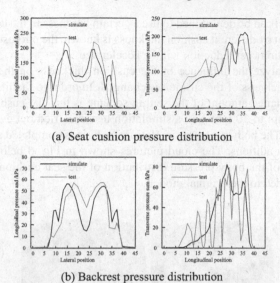

(a) Seat cushion pressure distribution

(b) Backrest pressure distribution

Fig. 5. Comparison between simulation and test of transverse and longitudinal pressure curves of seat cushion and backrest

Table 2. Comparison of body pressure indexes between simulation and test under H-point device

Region	Body pressure index	Test	Simulation	Error
Cushion	Peak pressure/kPa	12.24	12.68	3.59%
	Average pressure/kPa	4.21	3.90	−7.36%
	Contact area/cm^2	1093.55	1185.16	8.37%
Backrest	Peak pressure/kPa	9.31	8.92	−4.19%
	Average pressure/kPa	3.22	2.93	−9.01%
	Contact area/cm^2	595.48	643.8	8.11%

3.2 Analysis of Factors Affecting the Static Comfort of Ejection Seats

3.2.1 Influence of Cushion Thickness on Pressure Distribution of Cushion Body

The cushion thickness is an important factor affecting the riding comfort. Under certain density and hardness conditions, the thicker the cushion, the better the comfortable feeling of the human body will be. After actual seating, most of the weight of the human body is concentrated near the tubercle of the sitting bone, so the cushion under the buttocks should be thick enough to support the passengers. If the thickness is too small, the passengers will feel that the seat frame will collide when they are seated. However, if the cushion is too thick, the cost of the seat will increase. Therefore, it is necessary to consider a balance between the production cost and the comfort of the seat.In the early stage of cushion design, the cushion thickness at the corresponding position is usually determined according to the dummy layout and seat structure in the general layout. If the cushion product can be determined to use a certain hardness in the design stage, and the comfort brought by different cushion thickness is known, the purpose of seat product design and production cost reduction will be accelerated.

In order to analyze the influence of the cushion thickness on the body pressure distribution, the thickness of the cushion is changed. In order to facilitate analysis and observation, the change amount of the setting scheme is that the cushion thickness is increased from 5mm to 20 mm, and a simulation test is carried out every 5mm of the cushion thickness. The body pressure simulation analysis is completed according to the same simulation conditions. The cloud pictures shown in Fig. 6 below represent the simulation results of the body pressure distribution of the seat cushion when different thicknesses are added to the original model.

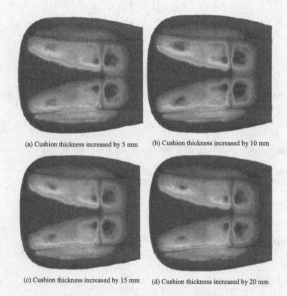

(a) Cushion thickness increased by 5 mm   (b) Cushion thickness increased by 10 mm

(c) Cushion thickness increased by 15 mm   (d) Cushion thickness increased by 20 mm

Fig. 6. Cloud Chart of Seat Cushion Body Pressure Distribution under Different Thickness

According to the form of pressure nephogram under different thickness of seat cushion added in Fig. 6, the concentration degree of larger pressure distribution in the hip area of seat cushion is greatly reduced, and there is no obvious change in contact area. The corresponding change trend chart is drawn according to the change of body pressure index in different regions, as shown in Fig. 7 below.

It can be seen from the trend chart that: on the original basis, with the increase of cushion thickness, the maximum pressure value is 1/2 in the hip area and 7/8 in the flank area, showing a downward trend; The average pressure value in the thigh area 3/4, the flank area 7/8 has a large degree of reduction, but in the hip area and the front of the seat cushion has no obvious change; The change of contact area in each area is not obvious. Therefore, the change of cushion thickness has a great impact on the maximum pressure and average pressure in some areas.

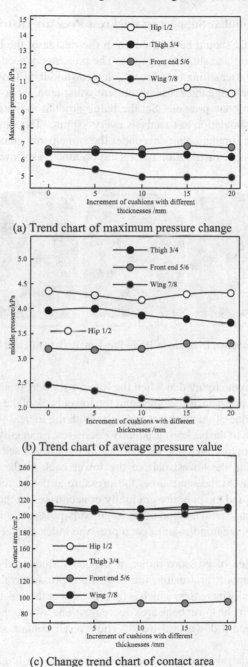

(a) Trend chart of maximum pressure change

(b) Trend chart of average pressure value

(c) Change trend chart of contact area

Fig. 7. Body Pressure Index Change Trend under Different Cushion Thickness

3.2.2   Influence of Lumbar Support on Backrest Pressure Distribution

The profile features that should be focused on in the design of the backrest profile are mainly the waist support and shoulder support. The passengers need to keep the natural curvature of the spine after sitting to reduce the fatigue and long-term discomfort of the waist. By changing the bulge thickness of the seat waist area, it shows the influence of the bulge thickness on body pressure. Set the bulge amount to increase from 0 mm to 30 mm, and conduct simulation test analysis every 10 mm. The simulation analysis of different waist support simulation models under the same parameters is carried out, and the backrest body pressure distribution results are obtained, as shown in Fig. 8.

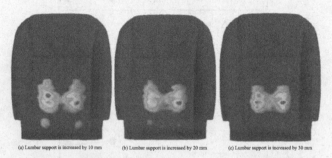

(a) Lumbar support is increased by 10 mm   (b) Lumbar support is increased by 20 mm   (c) Lumbar support is increased by 30 mm

Fig. 8. Cloud Distribution Diagram of Body Pressure under Different Lumbar Support

It can be seen from the figure that when the lumbar support is small, there is a large concentrated pressure between the backrest and the lower back of the passengers, and the maximum pressure value of the backrest is small. With the increase of lumbar support, the overall contact area of the backrest gradually decreases, especially the contact degree between the upper shoulder area and the bottom of the lower back decreases, and the contact pressure area at the lowest part of the lower back of the backrest gradually decreases to disappear. At the same time, the pressure at the lumbar support position continues to increase, and the pressure gradually concentrates on the lumbar position.

According to the statistical method of regional body pressure index and the regional pressure index change, the trend chart of each pressure index change is drawn as shown in Fig. 9 below.

Through the change of pressure index, it can be further analyzed that: with the increase of lumbar support protrusion, the maximum pressure value of 11/12 in the upper back area gradually increases, which is displayed near the waist, and the maximum pressure value of 9/10 in the lower back area decreases; The average pressure in the upper and lower back showed a downward trend; The overall contact area is significantly reduced.

The simulation analysis results show that increasing the lumbar protrusion can increase the maximum pressure value of the waist, and can also reduce the resistance pressure of the lower part of the backrest in contact with the human body due to the small lumbar support, but on the other hand, it will bring a certain reduction in shoulder comfort, because it will greatly reduce the contact area of the shoulder position.

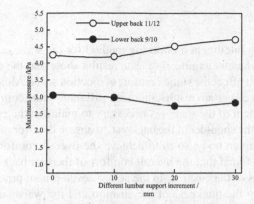

(a) Trend chart of maximum pressure change

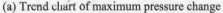

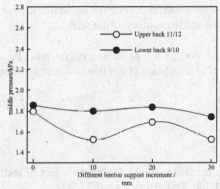

(b) Trend chart of average pressure value

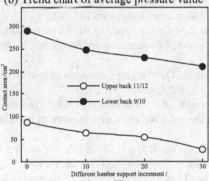

(c) Change trend chart of contact area

Fig. 9. Schematic diagram of index change trend in each area of backrest

4 Conclusion

This paper proposes an intelligent monitoring method for the static comfort of ejection seats based on human factors engineering. The results show that the cushion thickness and lumbar support will affect the static comfort of ejection seats. Although the research in this paper has achieved certain results, there are still many shortcomings. In the actual research and development of the seat, it is necessary to maintain the common matching between the waist and the shoulder of the backrest, to ensure the overall fit of the backrest and the back of the human body, so as to achieve the overall comfort of the backrest. In addition, this study found that the overall comfort of the seat back is closely related to the elastic properties of the cushion. In the actual development process of the seat, in addition to considering the thickness of the cushion and the waist support, we should also pay attention to the material and elastic modulus of the cushion. The appropriate material and appropriate elastic modulus can provide better support and buffering effect, so as to further improve the static comfort of the seat.

Aknowledgement. A school-level project of Beijing Polytechnic, Project Name: Research on layout design and static comfort simulation of ejection seat based on human factors engineering (KM202210858003).

References

1. Havelka, A., Nagy, L., Tunák, M., et al.: Testing the effect of textile materials on car seat comfort in real traffic. J. Ind. Text. **51**(5), 740–767 (2021)
2. Han, Y.H., Jia, Y.L., Li, W.L., et al.: Seat comfort evaluation based on psycho-physiological measurement method. Packag. Eng. **41**(6), 150–156 (2020)
3. Long, J., Guo, P.C., Chen, Z.M., et al.: Applying BP neural networks optimized with artificial bee colony algorithm to automobile seat comfort. Mech. Sci. Technol. Aerosp. Eng. **2**, 273–281 (2020)
4. Li, M., Gao, Z., Gao, F., et al.: Quantitative evaluation of vehicle seat driving comfort during short and long term driving. IEEE Access **99**, 1 (2020)
5. Zhang, H., Meng, L., Gong, Y., et al.: The influence of backrest angles on the passenger neck comfort during sleep in the economy class air seat without head support. Int. J. Ind. Ergon. **84**(3), 103074 (2021)
6. Moon, J.S., Tridib, K.K., Sung, B.H., et al.: Study on seating comfort of polyurethane multilayer seat cushions. Int. J. Automot. Technol. **21**(5), 101035 (2020)
7. Anjani, S., Song, Y., Hou, T., et al.: The effect of 17-inch-wide and 18-inch-wide airplane passenger seats on comfort. Int. J. Ind. Ergon. **82**(1), 103097 (2021)
8. Wegner, M., Martic, R., Franz, M., et al.: A system to measure seat-human interaction parameters which might be comfort relevant. Appl. Ergon. **84**, 103008 (2020)
9. Yadav, S.K., Huang, C., Mo, F., et al.: Analysis of seat cushion comfort by employing a finite element buttock model as a supplement to pressure measurement. Int. J. Ind. Ergon. **86**, 103211–103216 (2021)
10. Mondal, P., Arunachalam, S.: Finite element modelling of car seat with Hyperelastic and viscoelastic foam material properties to assess vertical vibration in terms of acceleration. Engineering **12**(3), 177–193 (2020)

11. Gao, K.Z., Luo, Q., Zhang, Z.F., et al.: Vibration comfort evaluation of vehicle seat based on body pressure distribution. Automot. Eng. **44**(12), 1936–1943 (2022)
12. Zhang, Z.F., Lu, X.H., Gao, K.Z., et al.: Discussion on comfort of vehicle seat cushion based on best-worst scaling method. J. Highway Transp. Res. Dev. **40**(2), 230–237 (2023)
13. Fan, Q.H., Jiang, X.C., Wu, X.L., et al.: Experimental research on optimization design and comfort of locomotive seat based on ideal pressure. J. Mech. Eng. **58**(10), 383–394 (2022)
14. Talhah, S.A., Noor, Z.K., Annayath, M., et al.: Ranking model for human seating comfort factors in automobiles: a best worst approach. Int. J. Serv. Oper. Manag. **42**(3), 353–378 (2022)
15. Vanacore, A., Lanzotti, A., Percuoco, C., et al.: A model-based approach for the analysis of aircraft seating comfort. Work **68**(s1), S251–S255 (2020)

Design of Mobile Education Evaluation System for College Students Based on Digital Badge Technology - Taking Legal Education as an Example

Yu Zhao[✉] and Liang Zhang

Changchun University of Finance and Economics, Changchun 130000, China
zhaoyy4100@163.com

Abstract. In the process of education evaluation, a lot of analysis and calculation are needed for different parameters. Once the original data is abnormal, the reliability of the evaluation results will be greatly reduced. Therefore, a design of mobile education evaluation system for college students based on digital badge technology is proposed - taking legal education as an example. In hardware design, FK-SK7-M2 is used as the storage device of the system to meet the data management requirements of the system, and STC12C5A60S2 is used as the digital terminal microprocessor of the system to meet the operation requirements of the evaluation system. In the aspect of software design, the digital badges with visual images are designed and classified according to the educational objectives. In the stage of education evaluation, the principal component analysis method is introduced, and with the help of an orthogonal transformation T, the original random digital badge vector whose components are related is transformed into a new random digital badge vector whose components are not related, and the contribution rate of the cumulative eigenvector in the covariance matrix is used to achieve the final teaching evaluation. In this test result, the evaluation result error of the design system for the sample data is stable within 0.15, and the average evaluation accuracy is 98.49%, with high reliability.

Keywords: Digital Badge Technology · Educational Evaluation · Digital Terminal Microprocessor · Principal Component Analysis · Random Number Badge Vector · Cumulative Eigenvector

1 Introduction

The evaluation of educational quality in colleges and universities is a relatively complex problem. The evaluation of educational quality includes many factors, such as educational conditions, curriculum difficulty, teacher education, and learning effects. They interact with each other. At the same time, the relationship between teachers and students is complex, and there are many factors that affect the quality of education [1]. At present,

L. Yun et al. (Eds.): ADHIP 2023, LNICST 550, pp. 20–35, 2024.
https://doi.org/10.1007/978-3-031-50552-2_2

there is no recognized and ideal educational quality evaluation system. As far as the current research situation is concerned, it focuses on three aspects: the first is the research on the evaluation subject, the second is the research on the content of the education quality evaluation system, and the third is the research on how to finally evaluate the education quality grading method after the indicators in the system are determined [2]. First of all, the study of the subject of education quality evaluation. There are many ways or methods to evaluate the quality of education, such as teacher self-evaluation, peer evaluation, administrative leadership evaluation, expert evaluation, and student evaluation of teachers [3]. Due to the different roles of evaluation subjects, their roles in evaluation should be different. Each evaluation method and its results are only a part of education quality evaluation, not the whole of education quality [4]. Due to the large number of college teachers and the frequent evaluation times, if the organization leaders and peer experts adopt the evaluation method of census, it will not only take time and effort, but also be difficult to operate because it is impossible to exclude the influence of interpersonal relationships, unfamiliar with the education process and other factors [5]. Therefore, the way of student oriented education quality evaluation of teachers is widely adopted by most colleges and universities. Since the 1980s, colleges and universities have gradually carried out student evaluation of teaching activities, which has played a certain role in promoting the quality of education in colleges and universities [6]. As the direct object of education, students have the right and ability to evaluate teachers' education. Due to the diversity of types of colleges and universities, the complexity of majors, and the uneven level of students, the requirements for teachers' education quality evaluation are also different [7]. The second is to determine the content of education quality evaluation. When designing the content of the education level evaluation system, it is difficult to quantify the role of a teacher in a certain course or learning stage because learning and development is a continuous process and the learning and growth environment is diverse [8]. Generally, curriculum performance is not the main indicator, or education effect is not the main indicator, The evaluation content is put on the education process [9]. From the perspective of process management, the school education process is manifested in the interaction of multiple factors and the combination of multiple links. It is also difficult to compare the education of different subjects, different courses, different education links, and different teaching objects. There is personal bias among evaluators in the process of educational evaluation, which makes the evaluation results not objective. In addition, inconsistent scoring standards may also lead to deviations in the evaluation results. Education evaluation usually relies on students' exam scores, neglecting their comprehensive abilities and potential, and this indicator evaluation cannot fully reflect their actual level and development. Therefore, the consideration of the education quality evaluation system is mainly designed from the most basic factors that can directly reflect the education level and have commonalities.

From the existing education level evaluation system, the design of indicators is mainly reflected in the following aspects. Education attitude, whether the education is serious and responsible, whether the class spirit is full, whether the lesson preparation is sufficient, and whether the coaching, question answering, and homework correction are serious. Whether the education content, content selection and processing are appropriate, whether the concept is accurate and clear, whether the key points are prominent, whether the difficulty and depth are appropriate, whether the theory is closely linked with the practice, and whether the content is rich. Whether the teaching ability and organization are clear, Whether the language is vivid, concise and attractive, whether the key points and difficulties are described accurately, and whether the blackboard writing is neat. Whether the education methods are individualized and flexible, whether they focus on inspiring students' innovative awareness and ability, and whether they focus on communication and interaction with students. To teach and educate people, whether they are rigorous and exemplary in their studies, and whether they are strict and fair in their requirements for students. Education effect: whether it can promote students to think positively, whether students' scores are improved, and whether students have a comprehensive grasp of knowledge points. Because different schools have different understanding and emphasis on education quality, there are certain differences in the content of evaluation. Analyze the current evaluation methods of education quality. After the establishment of each indicator system in the education quality evaluation system, certain methods should be used to process these data, to get the final education quality level. However, the corresponding evaluation methods have played a positive role in improving the quality of education and promoting the level of teacher education, but these methods have certain shortcomings. The traditional teaching quality evaluation method and fuzzy comprehensive evaluation method can no doubt get rid of the direct influence of human factors on the evaluation results. The Markov chain evaluation method only evaluates students' transcripts, which obviously has a great one-sided nature.

For this reason, this paper takes legal education as an example, proposes the research on the design of mobile education evaluation system for college students based on digital badge technology, and analyzes and verifies the operation performance of the design system through comparative testing. In terms of hardware, FK-SK7-M2 is used as the storage device of the system, and operating parameters are set according to its configuration. The microprocessor selected for the system is the STC12C5A60S2 chip, which has fast speed, low power consumption, and strong anti-interference ability, and can meet the operational requirements of the system. In terms of software, the digital badge designed using the evaluation gauge correlation model has strong objectivity and can provide reliable data support. The principal component analysis method is used to achieve comprehensive evaluation of mobile education for college students.

2 Design of Mobile Education Evaluation System for College Students

The mobile education evaluation system for college students includes hardware and software parts. The overall structure of the mobile education evaluation system for college students is shown in Fig. 1.

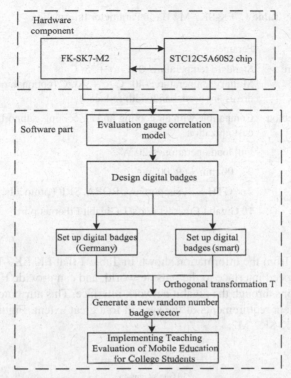

Fig. 1. Structure of Mobile Education Evaluation System for College Students

From Fig. 1, it can be seen that the hardware part mainly consists of FK-SK7-M2 and STC12C5A60S2 chips. The software part mainly designs digital badges through the evaluation gauge correlation model, setting up digital badges (German) and digital badges (intelligent) respectively. By using orthogonal transformation T, a new random number badge vector is generated to achieve the design of a mobile education teaching evaluation system for college students.

3   Hardware Design

3.1   Storage Device Design

Considering that the mobile education evaluation system for college students designed in this paper is based on digital badge technology, and the yin deficiency needs to analyze and mark large-scale data, FK-SK7-M2 is used as a storage device. In terms of structure, FK-SK7-M2 adopts a single board design and a 2-bay bit group Raid 0. In addition, FK-SK7-M2 also adopts the Fengke NVME storage architecture, which has the characteristics of miniaturization, low power consumption and standard eXFAT file system. The relevant basic parameter information is shown in Table 1.

Table 1. FK-SK7-M2 Basic Parameter Information

Index	Parameter
Working temperature	Standard temperature: 0 °C to +55 °C Military temperature: -40 °C to + 70 °C (optional, only some hard drives are used, bandwidth reduced)
Heat dissipation method	Compatible with air cooling and air cooling, supporting structural customization
Power waste	Full load operation ≤ 20 W
Dimensions	92. 00 mm x 69. 00 mm
Data interface	2 × GTH × 1, supports AURORA, SRIO protocols
Control interface	10 Gigabit Ethernet ports/1 Gigabit Ethernet port

It can be seen from the information shown in Table 1 that FK-SK7-M2 provides a 2-way 10G high-speed interface to the outside world, and can provide FTP or network disk access functions through the gigabit network interface. This attribute feature meets the data management requirements of the system to a great extent. Figure 2 shows the configuration of FK-SK7-M2.

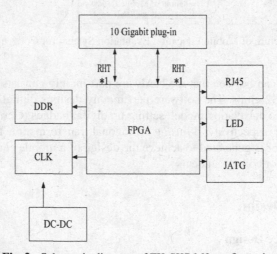

Fig. 2. Schematic diagram of FK-SK7-M2 configuration

In combination with the above configuration settings, the operation parameters of FK-SK7-M2 are shown in Table 2.

With the help of the configuration of operation parameters shown in Table 2, FK-SK7-M2 can be widely used in data acquisition, record storage and data management in related fields, and has high applicability to the mobile education evaluation system for college students designed in this paper.

Table 2. FK-SK7-M2 Operation Parameter Information

Index	Parameter
Storage capacity	Supports 2 NVMe SSDs with M. 2 interfaces; Capacity support for 2TB/4TB/8TB/16TB with multiple specifications available
Storage bandwidth	Continuously stable recording bandwidth $\geq$ 2GB/s; Continuously stable read bandwidth $\geq$2 GB/s
File management	Standard File System (exFAT File System)
Control protocol	The network adopts standard FTP communication protocol; Support standard FTP tools for accessing files
Download bandwidth	Gigabit Ethernet port $\geq$100 MB/s; 10 Gigabit Ethernet port $\geq$ 1000 MB/s
Software function	Real time recording function, data playback function, data access, online file management function, self inspection and fault detection function, abnormal fault tolerance protection mechanism, and other functions

3.2 Model Selection of Digital Terminal Microprocessor

Since cost control and energy consumption reduction should be considered in the project, the microprocessor itself is required to support low-power design with high integration, and include as many functional peripheral modules as possible, such as multi-channel I/O, analog-to-digital conversion, timer, UART serial communication, etc. required by the system. At the same time, it is necessary to respond to the sampled signals in a fast and real-time manner for processing and judgment, which has certain requirements on the processing speed of the microprocessor. The 1T enhanced series 51 single chip microcomputer produced by domestic STC company has great competitive advantages. The STC series 51 single chip microcomputer is compatible with 8051 instructions and pins, and its internal integrated large capacity memory is FLASH process. The FLASH ROM of STC12C5A60S2 single chip microcomputer is up to 60K. With this process memory, users can easily erase and rewrite instantaneously by using electricity. STC series MCU supports.

Serial program programming 1381. To sum up, STC series MCU has very low requirements for developing equipment, and the development time is also greatly shortened. At the same time, this series of MCU can encrypt the programs written into it, which also protects the labor achievements of developers. In comprehensive consideration, the microprocessor model selected for this project is STC12C5A60S2 chip. It is a new 8051 single chip computer produced by Hongjing Science and Technology, whose speed is 8–12 times faster than the traditional 8051 single chip computer. It has the advantages of high speed, low power consumption and super anti-interference. Internal integrated MAX810 special reset circuit, 2-way PWM, 8-way 10 bit high-speed AD conversion, for motor control, strong interference occasions. The STC12C5A60S2 single-chip microcomputer mainly has the following characteristics:

(1) High speed: 1 clock/machine cycle, enhanced 8051 core, 6~12 times faster than ordinary 8051.
(2) Wide voltage: 5.5~4. 0 V, 2.1 ~3. 6 V.
(3) Add the second reset function pin/P4. 6 (high reliable reset, adjustable reset threshold voltage, when the frequency is less than 12 MHz, it is unnecessary to add an external power down detection circuit/P4. 6, which can save the data into EEPROM in time when the power is down, and it is unnecessary to operate EEPROM during normal operation.
(4) Low power design: idle mode (can be awakened by any one interrupt).
(5) Low power consumption design: power-off mode (wake-up by external interrupt), which can support falling edge/rising edge and remote wake-up.

Pins supporting power down wake-up: P3. 2/INTO, P3. 3/INTI, T0/P3. 4, TI/P3. 5, RxD/P3. 0, P1. 3/CCP0 (or P4. 2/CCP0), P1. 4/CCP1 (or P4. 3/CCP1), EX_LVD/P4. 6.

(6) Operating frequency: 0 ~35 MHz, equivalent to common 8051: 0~420 MHz.
(7) Clock: optional external product or internal R/C oscillator, 8/1 6/32/40/48/56/60/62K byte on-chip Flash program memory is set when ISP downloads the programming user program, with more than 100000 erasures.
(8) 1280 byte on-chip RAM data memory.
(9) High capacity on-chip EEPROM function, ISP/AP with more than 100000 erasures, programmable in the system/programmable in the application, without programmer/emulator.
(10) 8-channel, 10 bit high-speed ADC, speed up to 250000 times/second, 2-way PIM can also be used as 2-way D/A.
(11) The 2-channel capture/comparison unit (CCP/PCAPWM) can also be used to implement 2 more timers or 2 external interrupts (support rising edge/falling edge interrupts).
(12) Two 16 bit timers (compatible with common 8051 timer T0/T1), and two more timers can be realized by 2-way PCA.
(13) Programmable clock output function (TO outputs clock at P3. 4, TI outputs clock at P3. 5, BRT outputs clock at P1. 0.

In addition, STC12C5A60S2 is also equipped with an independent baud rate generator and SPI high-speed synchronous serial communication interface dual serial ports. With the support of full duplex asynchronous serial port (UART), it is compatible with the common 8051 serial port. Time sharing multiplexing can be used as three groups. In terms of structure, STC12C5A60S2 is an advanced instruction set structure that is compatible with the common 8051 instruction set. It has hardware multiplication/division instructions. It uses the common I/O ports (36/40/44), and after reset, it is a quasi bidirectional port/weak pull-up (common 8051 traditional I/O ports). In the specific operation process, it can be set into four modes, quasi two-way port/weak pull-up, strong push pull/strong pull-up, which are only input/high resistance, so as to meet the operation needs of the mobile education evaluation system for college students in different states.

4   Software Design

4.1   Digital Badge Design

Digital badges, as digital stamps and marks in an online environment, facilitate timely feedback in classroom teaching. For this reason, this paper designs an intuitive digital badge. Figure 3 is the schematic diagram of the digital badge in the system.

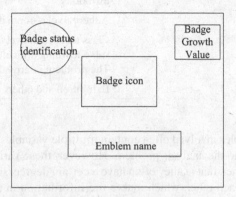

Fig. 3. Schematic diagram of digital badge

Class teaching time is very limited. Only the badges needed by teachers are presented. Teachers can find the badges they need faster to issue when evaluating, improve the efficiency of badge issuance, and complete the timely evaluation of classroom teaching. Students can also check their badges when logging into the evaluation system. Schematic diagram of the digital badge seen in the student port. Students can see the name of each badge and its corresponding growth value, the discipline in which it was obtained, and the specific time when it was obtained. The design of this evaluation system and the development team designed this system to develop a digital badge evaluation system for students in the field of comprehensive quality evaluation. This system involves a large number of evaluation systems, including.

The setting of character badges is also rich in variety, involving morality, intelligence, physique, beauty, labor and other aspects. However, the original digital badge in the system.

It is not necessarily appropriate for the needs of information technology classroom teaching. Therefore, this paper analyzes the preliminary proposed basis.

Based on the mobile education evaluation system of digital badges, combined with the correlation model of digital badges and evaluation gauges, 18 key badges are selected as the key research data of this study. The set digital badges are shown in Tables 3 and 4.

According to the way shown above, the digital badge is designed to provide a reliable data basis for subsequent education evaluation.

4.2   Analysis and Evaluation of Education Quality

Combined with the digital badges constructed above, this paper uses principal component analysis to analyze and evaluate the quality of education. When analyzing and processing

Table 3. Digital Badge Attribute Settings (Germany)

Emblem name	Badge Growth Value	Student activities
Attendance on time	1	Attend classes in the computer classroom on time
Bring books to class	1	Teaching with information technology textbooks
Discipline Pacesetter	1	Adhere to classroom discipline
Best Group	2	Class teacher, actively helping classmates solve learning difficulties The problem of arrival
Be ready to help others	1	Exhibition and others

teaching data, the samples involved often contain multiple variables, and more variables will bring complexity to the analysis problem. However, these variables have a certain dependence on each other, that is, they often have a certain degree, sometimes even quite high correlation with each other, which makes the information in the observation data overlap to a certain extent [10]. In this paper, the mechanism of principal component analysis can be simply stated as follows: with the help of an orthogonal transformation T, the original random number badge vector whose components are related is transformed into a new random number badge vector whose components are not related.

On the one hand, the original random number badge vector covariance matrix is transformed into a diagonal matrix, and on the other hand, the original coordinate system is transformed into a new orthogonal coordinate system, so that it points to the p orthogonal directions where the sample points are scattered most widely, and then the multi-dimensional variable system is reduced in dimension, so that it can be converted into a low dimensional variable system with a high precision. The covariance matrix V and the correlation coefficient matrix R are a measure of the degree of correlation between various components of the random number badge variable x, and contain rich information. From the perspective of refining information, people hope to transform them into diagonal matrices through an orthogonal transformation, and each component of the new random digital badge vector generated from them will become irrelevant.

Let the digital badge orthogonal matrix to be sought T by:

$$T = [t_1, t_2, ..., t_p]_{t*t} \tag{1}$$

Among them, $t_j = [t_{1j}, t_{2j}, ..., t_{pj}]^T$, and j = 1,2,.., p.
Generated new random number badge vector u by:

$$u = [u_1, u_2, ..., u_p]^T \tag{2}$$

Then there is:

$$u = T^T x \tag{3}$$

Table 4. Digital Badge Attribute Settings (Smart)

Emblem name	Badge Growth Value	Student activities
Interactive experts	1	Actively raise your hand to answer questions
question answering	1	Answer the teacher's questions correctly
A knowledgeable listener	1	Listen attentively to the class
Finish one's homework	1	Complete learning tasks
Excellent homework	2	High quality completion of assignments (integrity, creativity Artistry, etc.)
Rich imagination	2	Proficient in thinking from multiple perspectives and imaginative
Best Ideas	1	Works with high originality and creativity
Technical experts	1	High level of professionalism in the work, proficient in operational skills Be good at discovering the excellence of classmates' works and expressing oneself
Commenting experts	1	Proficient in mastering the knowledge learned, able to transfer and expand knowledge
Be ready to help others	1	Exhibition and others
Knowledge expert	1	Significant progress in learning performance
Star of Progress	1	Actively participate in learning activities, group activities, etc
Actively participate in	1	Mutual assistance and sharing among team members for common progress
Extracurricular expansion	1	Extend classroom learning beyond the classroom and apply what is learned

I. e:

$$u_i = t_i^T x, i = 1, 2, ..., p \tag{4}$$

Set up:

$$X = [x_1, x_2, ..., x_p]^T \tag{5}$$

And formula (5) is a p-dimensional random vector, whose i-th principal component can be expressed as $u_i = t_i^T x$, $i = 1, 2, ..., p$, where t is the i-th column vector of orthogonal matrix T, and meets the following conditions:

(1) U1 is the random variable with the largest variance in formula (4);
(2) U2 is the random variable with the largest variance among the other variables unrelated to u1in formula (4);
(3) Uk is the same as u1, u2,The random variable with the largest variance among the other variables that are not related to uk-1,

k = 3, 4,…, p. On this basis, in the covariance matrix of the random number badge vector, the corresponding feature vector is a, and the cumulative contribution rate of the feature vector is the final teaching evaluation result, which can be expressed as.

$$A = (\sum a_i)(\sum a_i)^{-1} \qquad (6)$$

Among them, A It represents the final teaching evaluation result.

The evaluation of mobile teaching can be realized according to the way shown above.

5   Test Experiment Analysis

5.1   Test Environment Parameter Setting

In the testing phase, comparative testing is carried out based on the actual data information of law education. A total of 300 groups of data were determined to participate in the cps were used for training, and the remaining 30 groups of data were used for testing in order to test the performance of the system. The evaluation index parameters of the test data are shown in Table 5.

Table 5. Test data evaluation index parameters

Number/Group	Evaluation index parameters				
	Student behavior	Student feedback	Academic record	...	Teaching quality
1	0.81	0.59	0.75	...	0.76
2	0.35	0.71	0.73	...	0.53
3	0.86	0.92	0.78	...	0.85
4	0.83	0.49	0.77	...	0.74
5	0.69	0.65	0.28	...	0.62
...	...	...	...	...	...
30	0.75	0.63	0.74	...	0.74

On this basis, the system designed in this paper is used for testing. Through comparing the test results of different systems, its performance is analyzed.

5.2 Test Results

(1) System performance test.

The simulation experiment of teaching quality evaluation is realized through MAT-LAB 2013b programming, and the mean square error change diagram of the output results of the design system in this paper is shown in Fig. 4, the evaluation accuracy is shown in Fig. 5, the error square sum of the output results of the design system in this paper is shown in Fig. 6, and the fitness function curve of the output results of the design system in this paper is shown in Fig. 7.

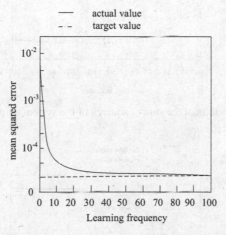

Fig. 4. Mean Square Error Variation Diagram of System Output Results

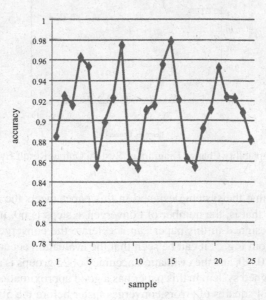

Fig. 5. Accuracy change diagram of system output results

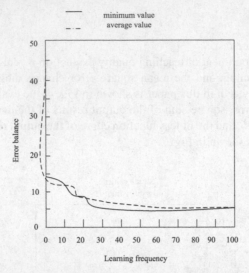

Fig. 6. Variation of the sum of squares of the system output error

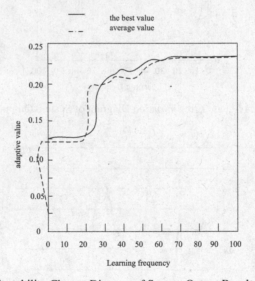

Fig. 7. Adaptability Change Diagram of System Output Result Evaluation

Figure 4 shows that the system designed in this paper meets the stop condition in the 60th generation, that is, the number of convergence steps is 60, indicating that the use of the system designed in this paper can accelerate the convergence speed in the evaluation phase. From Fig. 5, it can be seen that the evaluation accuracy of 29 groups of 30 test samples is 0. 90, and the evaluation accuracy of 22 groups is above 0. 87, It can be seen that the designed system in this paper has a good approximation effect. Figure 6 shows that the sum of squares of errors converges faster before the 5th generation, and the convergence speed is relatively slow in the 10th to 30th generation. After the number of iterations is 34, the sum of squares of errors of the network is stable, indicating that the designed system in this paper can achieve global optimization faster. Figure 7 shows that the fitness of the output results of the design system in this paper converged quickly before the 10th generation, and basically reached a stable state after 45 iterations. It can be seen that the adaptability of the system is high. In a word, the internal mechanism of the design system in this paper determines its training and evaluation performance. From the perspective of evaluation accuracy and adaptability, the design system in this paper is effective and robust.

(2) Comparative test.

In order to verify the performance of this model, the BP engine is used.

The sample data is simulated by network method, and the comparison test is carried out. The absolute error comparison diagram of evaluation results is shown in Fig. 8.

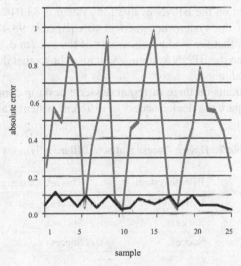

Fig. 8. Comparison Diagram of Absolute Error of Evaluation Results

Combined with the comparative test results in Fig. 8, it can be seen that the absolute error of the evaluation of the teaching quality evaluation system based on BPNN is distributed between 0 and 0. 98, and the absolute error of 10% sample evaluation is relatively large. The absolute error range of the system evaluation designed in this paper is 0–0. 15, and the absolute error of 90% sample evaluation results can be controlled within 0. 2. The prediction results of individual samples of BPNN system deviate greatly, and the evaluation results of this design system are relatively stable. Compared with BPNN teaching quality evaluation system, it can be seen that the evaluation system designed in this paper has better performance.

The comparison results of evaluation accuracy of different systems are shown in Table 6.

Table 6. Comparison of Test Result Precision/%

Test system	BPNN system	This article designs a system
Average evaluation accuracy	83.04	98.49
Maximum evaluation accuracy	92.44	99.30
Minimum evaluation accuracy	80.43	95.63

According to the test results shown in Table 6, the average evaluation accuracy of 30 groups of data based on the BPNN evaluation system is 83.04%, while the average evaluation accuracy of the system designed in this paper is 98.49%, an increase of 15.45%. It can be seen that the evaluation results of the system designed in this paper are obviously better than the BPNN system, and it also shows that the system is feasible for teaching quality evaluation.

Access the two systems on three different devices: desktop computers, tablets, and smartphones, and compare the device access of different systems, as shown in Table 7.

Table 7. Device Access Status of Different Systems

Equipment type/type	BPNN system	This article designs a system
Desktop computer	Fail	Success
Ipad	Success	Success
Smartphone	Success	Success

From Table 7, it can be seen that the BPNN system does not support desktop access, while this article designs a system can be successfully accessed on all devices, indicating that this article designs a system has strong inclusiveness.

6 Conclusion

Education is a dynamic process that integrates teaching and learning. There are many factors that affect the quality of education. In addition, these factors have different degrees of influence. Therefore, it is difficult to express the evaluation results with a mathematical analytic formula. In fact, it is a relatively complex, nonlinear comprehensive decision-making problem. Most evaluation methods comprehensively evaluate the quality of education, and it is difficult to exclude all kinds of randomness and subjectivity, It is easy to cause distortion and deviation of evaluation results, and there is irrationality. For this reason, this paper proposes the design of mobile education evaluation system for college students based on digital badge technology - taking legal education as an example, which effectively ensures that the design system can meet the objective application requirements under different conditions, and has high reliability. The storage device of the system is selected as FK-SK7-M2, and the microprocessor is selected as STC12C5A60S2 chip, ensuring that the system has a faster operating speed and lower power consumption to meet the system's operational requirements. In terms of software, digital badges and principal component analysis are combined to achieve educational evaluation. The future system can expand the dimensions of evaluation indicators to more comprehensively evaluate students' comprehensive qualities and abilities.

References

1. Ma, X.N., Zhao, Z.F.: Aspect-based sentiment analysis model based on neural network. Comput. Simul. **39**(11), 5 (2022)
2. Zhao, Y.: Teaching traditional Yao dance in the digital environment: Forms of managing subcultural forms of cultural capital in the practice of local creative industries. Technol. Soc. **69**(C), 101943 (2022)
3. He, S., Wu, Y.: Discussion on the application of computer digital technology in the protection of intangible cultural heritage. J. Phys. Conf. Ser. **1915**(3), 032048 (2021)
4. Li, F., Wang, Z.: Application of digital media interactive technology in post-production of film and television animation. J. Phys. Conf. Ser. **1966**(1), 012039 (7pp) (2021)
5. Donnelly, R., Maguire, T.: Building digital capacity for higher education teachers: recognising professional development through a national peer triad digital badge ecosystem. Euro. J. Open Distance E-Learn. **23**(2), 1–19 (2021)
6. Gengxuan, C., Qinmin, J., Hao, L.: Rethinking the rise of global central bank digital currencies: a policy perspective. Contemp. Soc. Sci. **8**(1), 16 (2023)
7. Govindarajoo, G., Lee, J.Y., Emenike, M.E.: Proof of concept for a thin-layer chromatography digital badge assignment within a laboratory practical exam for a nonchemistry majors' organic chemistry lab. J. Chem. Educ. **2021**(9), 98 (2021)
8. Vc, A., Gn, A., Rks, A.: The future of continuous learning–Digital badge and microcredential system using blockchain - ScienceDirect. Global Trans. Proc. **2**(2), 355–361 (2021)
9. Bughin, J., Kretschmer, T., Zeebroeck, N.V.: Digital technology adoption drives strategic renewal for successful digital transformation. IEEE Eng. Manag. Rev. Reprint J. Eng. Manager **49**(3), 103–108 (2021)
10. Liu, S., He, T., Li, J., et al.: An effective learning evaluation method based on text data with real-time attribution - a case study for mathematical class with students of junior middle school in China. ACM Trans. Asian Low-Resour. Lang. Inf. Process. **22**(3), 63 (2023)

A Method of Resolving the Conflict of Shared Resources in Online Teaching of Design Professional Artworks Based on Feedback Integration

Bomei Tan$^{(\boxtimes)}$ and Rong Yu

Nanning University, Nanning 530200, China
tanbomei861226@163.com, yyrr22@yeah.net

Abstract. In order to solve the problem of massive online teaching shared resources conflict, reduce the error rate of shared resources conflict resolution, and improve the success rate of conflict resolution, the feedback integration concept was introduced. Taking the online teaching shared resources of design professional artworks as an example, the research on the conflict resolution method of online teaching shared resources of design professional artworks based on feedback integration was carried out. First, a strategy model for resolving the conflict of shared resources in online teaching is established. Secondly, the control protocol of teaching resource sharing collaborative service is designed to achieve the goal of intelligent interconnection, resource sharing and collaborative service between online teaching shared resources. Using the feedback integration principle, design an online teaching sharing resource integration and collaboration platform, and integrate the online teaching sharing resources of design professional artworks. On this basis, according to the first come first served resource conflict resolution strategy, resolve the design professional art online teaching sharing resource conflict. The experimental analysis results show that after the application of the new method, the error rate of online teaching shared resource conflict resolution is low, up to 0.2%, and the application effect has significant advantages.

Keywords: Feedback Integration · Design Discipline · Artwork · Online teaching · Sharing Resources · Conflict · Digestion

1 Introduction

Teaching resources are defined in the Education Dictionary: instructional resources refer to various resources that support teaching activities [1]. Teaching resources mainly come from two aspects: on the one hand, they are the original available resources in the real world, and on the other hand, they are resources designed specifically for learning purposes [2].

In recent years, the construction and development of the network platform have achieved remarkable results, playing an important role in promoting student sharing and

L. Yun et al. (Eds.): ADHIP 2023, LNICST 550, pp. 36–52, 2024.
https://doi.org/10.1007/978-3-031-50552-2_3

improving the quality of education and teaching. However, due to various reasons, there are problems that can not be ignored and need to be solved urgently in the concept, capital, resources, technology, mechanism and online learning of platform construction. Only through the combination of resource construction and continuous updating, platform construction and technology upgrading, characteristic resources and students' needs, network platform and classroom teaching, the implementation of long-term mechanism, the development of "effective measures" to promote the sustainable development of platform construction, and the realization of full opening of resources and "large area" sharing of students [3].

With the development of education, groups of young teachers have stepped onto the platform of colleges and universities. However, most young teachers lack teaching experience, which leads to poor classroom teaching results and can not meet the expectations of schools and students [4]. The effect of college teachers' classroom teaching has been paid more and more attention by students, teachers and educators. College classroom teaching is not only a simple explanation, but also pays more attention to the effect of explanation and students' welcome to the curriculum [5]. Therefore, how to strengthen the teaching practice ability of young teachers and improve the classroom teaching effect has become an urgent problem for higher education. At present, domestic researchers and educators have proposed various methods, one of which is that teachers can share teaching related resources in real time through online teaching sharing platform. The sharing of teaching resources refers to breaking the original boundaries of teaching resources owned by different schools, different departments or individuals through mature computer technology and modern network technology under the guidance of relevant education departments within a certain area, and implementing the way of shared enjoyment through paid or free ways [6].

China has done a lot of theoretical research and practical exploration on the road of education informatization, and has made considerable progress, such as electronic book bags, distance education, etc. However, we should also see that there are still serious problems of "information island", reconstruction of teaching resources and low utilization rate in China's education at this stage. Using modern information technology to realize reasonable and efficient organization and management of the rapidly growing mass of teaching resources, so that high-quality teaching resources can be orderly co built and shared within a certain region, is an effective measure and way to solve the problems of "information islands" and information reconstruction, and is also one of the important ways to promote China's education informatization, It is of great significance and value to realize educational equity.

However, due to the huge scale of online teaching shared resources, in the actual operation process, resource conflicts, resource overlaps and other issues often occur, which seriously affect the utilization rate of online teaching shared resources, which is not conducive to the sustainable development of resource sharing, and needs to be addressed by scientific conflict resolution methods. The traditional resource conflict resolution method has a high error rate in the practical application process, which can not effectively resolve the resource conflict. Feedback integration can improve this problem, make full use of the output information of different platforms, add feedback links, make the platform from open loop to closed loop, and complete the task of identifying shared resources in online teaching. Based on this, this paper takes the online teaching and sharing resources of design professional artworks as an example, introduces the concept of feedback integration, and carries out the research on the conflict resolution method of online teaching and sharing resources of design professional artworks based on feedback integration. Firstly, a conflict resolution strategy model for online teaching shared resources was established, which can help solve conflict problems. Secondly, a collaborative service control protocol for teaching resource sharing was designed, which can coordinate the sharing and collaboration of multiple teaching resources, ensuring the smooth progress of the teaching process. Then, based on feedback integration, a shared resource integration and collaboration platform was designed, which can help teachers and students better share and collaborate, and improve teaching effectiveness. Finally, through the method of resolving conflicts in teaching shared resources, possible conflicts that may arise during the teaching process have been resolved, and the quality and efficiency of teaching have been improved.

2 Design of Conflict Resolution Method for Online Teaching Sharing Resources of Design Professional Artworks

2.1 Establish the Conflict Resolution Strategy Model of Online Teaching Shared Resources

The online teaching shared resource conflict resolution strategy model can effectively solve the problem of collaborative work between roles and parallel task resource conflict in the platform, which lays a good foundation for subsequent research.

The conflict resolution strategy model of shared resources for online teaching of design professional artworks established in this paper is shown in Fig. 1.

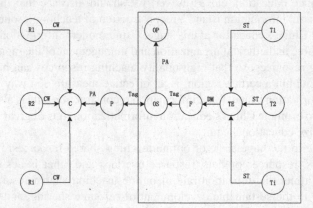

Fig. 1. Conflict resolution strategy model of online teaching shared resources

The model content in Fig. 1 is defined in detail as follows:

(1) R (role set) refers to the attributes of a role, including both the permissions that the role should have and the collaboration between roles.
(2) C (collaborative working set) refers to a set of multiple different roles working together to complete the same task. In the online teaching resource sharing platform, teachers and staff work together to complete teaching resource sharing [7].
(3) T (task set) refers to the attributes of each task in the system, including the relationship between tasks and roles.
(4) OP (operation set) refers to the minimum action set that can perform a certain function. In resource sharing, query, modify, and cancel the reservation object.
(5) OS (shared resource set) refers to the resources shared by parallel tasks. An operation on the shared time and database in the shared platform.
(6) TE (resource application time set) refers to the time when parallel tasks apply for the use of shared resources.
(7) F (Resource Resolution Policy Set) refers to the attributes of shared resource conflict resolution policies. If T has the right to use shared resources $F = (T, OS, OP)$, it means that T can perform OP operation on OS.
(8) P (permission set) refers to the mark of operation on an object. If R has permission $P = (T, OP)$, then R can perform OP operation on T.

In the model, the application of role collaboration and shared resource conflict resolution strategy greatly improves the working mode of the traditional sharing platform, and improves the reliability and accuracy of the sharing platform.

2.2 Design a Collaborative Service Control Protocol for Teaching Resource Sharing

After completing the establishment of the above conflict resolution strategy model for online teaching shared resources, next, design the control protocol for collaborative service of teaching resource sharing to achieve the goal of intelligent interconnection, resource sharing and collaborative service between online teaching shared resources. This paper adopts GCCP, the general control basic protocol, and its hierarchy diagram is shown in Fig. 2.

As shown in Fig. 2, GCCP processing unit completes the general control function of teaching resource sharing of vocational and technical education, and also has it to handle the transmission of shared application data in the network. Point to point and point to multipoint transmission of shared data is completed by the physical layer, data link layer and network layer of GCCP protocol layer. The network protocol in line with the national standard bears the collaborative services between the physical layer, data link layer and network layer [8] for the sharing of teaching resources in vocational and technical education. Among them, the physical layer is the lowest layer of the protocol, responsible for handling direct communication with physical media. In the collaborative service of teaching resource sharing, the task of the physical layer is to convert digital signals into suitable physical signals for transmission, and control the transmission rate and distance of data. The physical layer is also responsible for handling possible errors and interference during the transmission process, ensuring the reliable

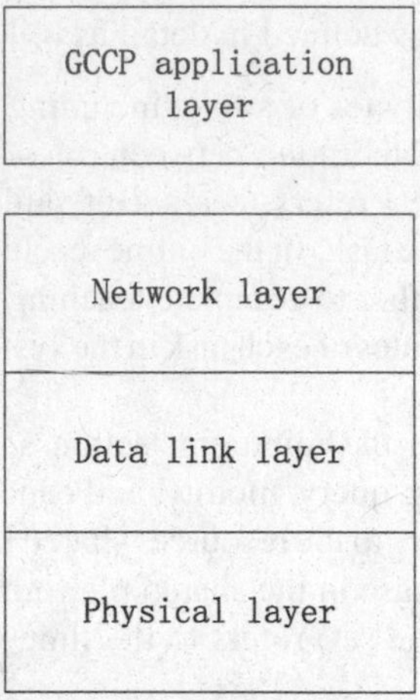

```
┌─────────────────────┐
│   GCCP application  │
│        layer        │
└─────────────────────┘

┌─────────────────────┐
│    Network layer    │
│                     │
├─────────────────────┤
│   Data link layer   │
│                     │
├─────────────────────┤
│   Physical layer    │
│                     │
└─────────────────────┘
```

Fig. 2. Hierarchy Diagram of GCCP Collaborative Service Control Protocol

transmission of data. Data link layer: The Data link layer is built on the physical layer and is responsible for dividing the data transmitted by the physical layer into data frames, and providing error detection and correction mechanisms. In the collaborative service of teaching resource sharing, the task of the Data link layer is to ensure the integrity and reliability of data, and detect and correct possible errors in the transmission process through checksum and other mechanisms. Network layer: The network layer is the highest layer of the protocol, responsible for implementing data routing and transmission. In the collaborative service of teaching resource sharing, the task of the network layer is to route data from the sending end to the receiving end, and achieve efficient data transmission by selecting the best path and using appropriate routing protocols. The network layer is also responsible for handling issues such as network topology, address allocation, and packet segmentation and reassembly. In general, the physical layer, Data link layer and network layer constitute the hierarchical structure of GCCP protocol. Each level has specific functions and tasks, and through collaboration and interaction between layers, efficient operation of collaborative services for teaching resource sharing and reliable data transmission are achieved.

In the general control GCCP network for sharing teaching resources of vocational skills education, all protocols involved in realizing interaction are shown in Fig. 2.2, and are completed by device A and device B. The GCCP network for sharing teaching resources of vocational and technical education established consists of core components such as general equipment, controller, configurator and gateway.

General equipment: It is generally the general control equipment of the network shared resource platform.

Controller: the equipment in the shared resource platform that operates and controls general network equipment, such as centralized controller, intelligent control terminal, etc.

Configurator: This configurator completes the configuration of other devices in the shared resource platform network, and assigns Network D and Device D to other devices.

Gateway: It connects two different networks to realize intelligent interconnection, resource sharing and collaborative services of devices in different networks. This core component can connect the IGRSIP main network and GCCP network, or connect two GCCP shared networks. In the shared resource platform network, a configurator is used to configure the functions of all devices, and the configurator configures independent network D and device ID [9] for each device in the network. The GCCP interaction involves protocol schematic diagram, as shown in Fig. 3.

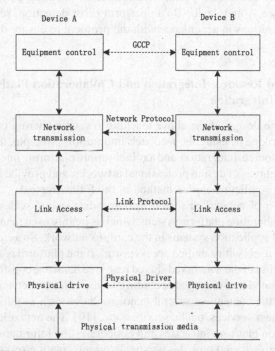

Fig. 3. Schematic diagram of GCCP interaction protocol

As shown in Fig. 3, the network ID consists of two octets, which are used to distinguish different network resources of the shared resource platform and are managed uniformly by the core component - configurator. GCCP interaction involves the protocol used for communication and data exchange between device A and device B. The specific working principle is: (1) Establishing a connection: Before communication between device A and device B, a connection needs to be established through the protocol. This can be achieved through a handshake process, where device A and device B exchange specific messages to confirm each other's identity and communication parameters. Once the connection is established, device A and device B can start transmitting data to each other. (2) Data exchange: Once the connection is established, device A and device B can use the GCCP protocol for data exchange. Device A can send requests or instructions to device B, which then performs corresponding operations based on the received requests or instructions and returns the required response or data to device A. This data exchange may include the transmission of teaching resources, updates of status information, etc.

(3) Message encapsulation and parsing: The GCCP protocol defines the format and structure of messages, and devices A and B need to encapsulate and parse messages in the manner specified by the protocol during data exchange. The encapsulation process combines data and control information into a specific message format for easy transmission and recognition. The parsing process parses the received message, extracting data and control information for use by the device. (4) Error handling and recovery: During data exchange, various error situations may occur, such as data loss, transmission errors, timeouts, etc. The GCCP protocol defines error handling and recovery mechanisms, where device A and device B can perform error detection, retransmission, or other necessary operations in accordance with the protocol to ensure data integrity and reliability.

2.3 Design Shared Resource Integration and Collaboration Platform Based on Feedback Integration

After completing the design of the above teaching resource sharing collaborative service control protocol, next, using the feedback integration principle, design the online teaching shared resource integration and collaboration platform, integrate the online teaching shared resources of design professional artworks, and provide basic support for the shared resource conflict resolution method in the following text.

The construction of networked shared resource database provides an information integration and collaboration platform system, which is mainly used to integrate the existing information and application systems in the campus network. So as to achieve unified control and provide users with a unified access portal. If the platform is supplemented by the campus application of the networked shared resource database construction platform, it is currently integrated and managed according to the existing network resources and campus needs, and the contents of several major modules such as information, search, collaboration, business services, public services, etc. [10]. The network shared resource database construction platform integrates and cooperates the information in the school, combs and summarizes it, and provides personalized information services for users.

The construction of information integration and collaboration platform mainly includes the following aspects:

1. Information integration platform and group management.

(1) Multiple integration schemes such as RSS, IFRAME, WebClipping and URL are provided, which can be integrated with campus and off campus websites, education, academic websites, blogs, network storage applications and other systems.
(2) Provide development level development framework and functional components to meet the requirements of functional expansion and application development; Provides functional components for portlet development on the portal.
(3) It provides the interaction function between user groups, and enables users in the group to manage their access to information. It mainly includes: user group selection, group customization, group free combination, etc.

2. Document storage and sharing services. Documents that provide designated users to browse or manage their own shared files are stored in the sharing service. Sharing is only the sharing of a file or folder. It is a temporary exchange of files between users.Users

can comment on files shared by others, and also view and manage the comment information of files shared by themselves. At the same time, it provides fast search and advanced search, and can search according to different file ranges. The shared platform system provides the user with a client backup tool, which allows the working directory in the user's machine to always be bound to a space in the network storage, so that files can be backed up within a settable time period. The backup function requires the following:

3. Personal collaboration application. Provides functions such as classification, publishing and browsing of notice announcements, and supports WYSIWYG online information editing; The system also supports the top setting, isolation and other functions of notification announcements. It can also be integrated with any module that needs to be reminded to provide centralized reminders and send and receive user messages.

4. Data integration. The integrated central database mode of the data integration platform provides a database mode that meets the needs of school data sharing and exchange. The developed integration interface provides interface support for different types of data sources, including: supporting data integration of mainstream RDBMS, such as Oracle, DB2, Sybase, SQL Server, and InfoMix; Support data integration of non mainstream RDBMSs, such as MySQL, Derby, HypersonicSQL, and PostgreSQL; Support the data integration interface of ODBC data sources such as Foxpro, Access, Excel, etc.; Support data integration of message type data sources such as JMS Queue and JMS Topic; Support file data integration such as formatted txt and XML; Support data integration of WebService; It supports other special types of data formats, such as data integration of LOB fields (BLOB, CLOB).

5. Data integration KM, topology management tools.

(1) Data integration KM module.

Provide data integration KM library, including more than 100 development packages for various data integration requirements.

(2) Topology management tools.

Manage data sources and scheduling agents, and support various data source interfaces such as RDBMS, text, message, and WebService.

3. Integrated design tools, integrated view tools, and integrated scheduling tools.

(1) Integrated design tools.

Provide graphical interfaces for design and development of data integration projects.

(2) Integrated viewing tool.

View the operation of the data integration project and debug the integration process.

(3) Integrated scheduling tool.

Schedule and control each data integration synchronization task to complete the customized data integration process.

2.4 Conflict Resolution of Teaching Shared Resources

After the above design of shared resource integration and collaboration platform based on feedback integration is completed, the next step is to resolve the conflict of shared resources in online teaching of design professional artworks.

First, allocate online teaching resource samples. According to the analysis of the deficiency of DPC algorithm, the improved algorithm adds a constraint in the sample allocation. MG-DPC algorithm first refers to DPC algorithm to calculate distance δ

And local density ρ. Then find the matching cluster center according to the decision graph heuristic. Other samples of the dataset i Samples with high local density and close distance j In the class cluster of, and the sample i, j They must be close neighbors to each other. If sample i, j If it is not a close neighbor, give the sample i Assign a new class label and mark it as "negative class". Other samples can be allocated to samples i In the negative class. When each sample is allocated, there may be some negative classes, which need to be merged into positive classes. Define 4 cluster boundaries: the definition expression of cluster boundaries is:

$$edge(A_p, B_N) = \{i | j \in MNN(i), j \in B_N, i \in A_p\} \tag{1}$$

Among them, A, B Is a class label, A_p Represent class cluster A Is a positive class, B_N Represent class cluster B Is a negative class, $MNN(i)$ Is a sample i The set of neighbors of. Any platform is composed of resources with different characteristics and categories, of which physical resources are easy to copy and purchase in the market, usually in materialized form. To realize the integration of shared resources, certain tools are necessary. In the shared resource conflict resolution method designed in this paper, shared resource integration technology is used to resolve resource conflicts. The integration technology includes mapping, topology and neural network. This paper chooses mapping technology as a means of conflict resolution. In the field of mathematics, mapping refers to the special correspondence between one set element and another set element. Different mapping definitions are generated according to different mapping objectives. Although the definitions of mapping are different, they are essentially the same, such as functions and operators. It should be noted here that the mapping between two number sets is a function, and the other mapping is not a function. The mapping can be defined as follows.

$$m : \langle e_s, e_t, r, k \rangle \tag{2}$$

Among them, e_s, e_t Represent the entities corresponding to the online teaching shared resources in the platform respectively; r express e_s And e_t Relations between them, such as equivalence, inclusion, overlap, etc.; k express e_s And e_t The closer its value is to 1, the more similar the two entities are. Through the mapping function of the integration unit, the shared resources of online teaching will be transformed into the final output information, and the transformation and utilization of shared resource elements will be realized.

When system resource conflicts occur in parallel tasks, they are handled according to the first come, first served resource conflict resolution strategy. Implement $Task_i(T_i)$ The agent of the task is assumed to be $Agent_i(A_i)$, Execute $Task_j(T_j)$ The agent of the task is assumed to be $Agent_j(A_j)$, $Agent_i$ And $Agent_j$'s application time is te_i And te_j.

Define 1 Parallel Task Set $T = \{T_1, T_2, ..., T_i\}$, where $i \in N = \{1, 2, ..., n\}$.

Define 2 Parallel Task Execution Set $A = \{A_1, A_2, ..., A_i\}$, where $A_i \in T = \{T_1, T_2, ..., T_i\}$.

Define 3 Request Time Set $te_i = \{te_1, te_2, ..., te_i\}$, where $i \in N = \{1, 2, ..., n\}$.

Execute Task Set $A = \{A_1, A_2, ..., A_i\}$ The decision objective function of is:

$$F(te_i, te_j) = \begin{cases} A_j, te_i \geq te_j \\ A_i, te_i < te_j \end{cases}, i, j \in N = \{1, 2, ..., n\} \tag{3}$$

When A_i And A_j When applying for online teaching sharing resources on the same day and in the same session, if $te_i \prec te_j$, online teaching shared resources are allocated to A_i, i.e. T_i; if $te_i \geq te_j$, the online teaching shared resources will be allocated to A_j, i.e. T_j. When the online teaching shared resources are assigned to an agent for execution, the online teaching shared resources of this session are immediately locked. Once locked, other parallel tasks cannot share the resources of this session, instead, they can reselect other sessions for resource sharing, or wait until the shared resources are unlocked, so as to effectively achieve the goal of shared resource conflict resolution.

3 Experimental Analysis

3.1 Experiment Preparation

The above content is the whole design process of the online teaching and sharing resource conflict resolution method of design professional artworks based on feedback integration proposed in this paper. Before the proposed conflict resolution method is put into practical use, the feasibility and resolution effect of the method need to be objectively tested. Based on this, the experimental analysis is carried out as shown below.

First of all, according to the method and content of the above design, build a highly adaptable online teaching shared resource conflict resolution development and testing environment. The specific requirements of the computer hardware operating environment required for the development of the component sharing platform and the computer software operating environment required for the development of the sharing platform are shown in Table 1.

According to the technical requirements and configuration shown in Table 1, build the development and test environment required for this experiment. In the experimental test, once the knowledge points, steps, organization forms and other contents required for the development of the entire shared resource curriculum are determined, the design and development work specific to a single shared resource can be started. Curriculum shared resources mainly include script development of teacher files and web page sharing content that currently exists in the Internet network. In order to effectively integrate resources into the database of the network shared resources platform, these curriculum resources need to be standardized by SCORM standards. The implementation process is shown in Fig. 4.

Script development and testing of basic data of curriculum shared resources are the data basis for processing shared application business on the platform of networked shared education and teaching resources. The functions related to the shared resource application business in all terminal applications on the education and teaching management platform will be based on the basic data, and the shared application processing will be completed on this basis. The basic data on the enterprise marketing platform has a huge amount of data. For end users with different functions, there are also certain operational permissions for basic data. For information security reasons, end users have the smallest set of basic data visual data under functional permissions.

Table 1. Computer Software Running Environment Settings

Build environment	Name	Technical requirement
Shared resource development environment	CPU	Dominant frequency above 1.7 GHz
	Memory	Above 2 GB
	Hard disk	Above 200 G
	Sound card	Provide (play gallery)
	Monitor	Widescreen display above 15 inches is recommended
	Provide Internet access conditions	The network speed provided shall be more than 100M for wired network card and more than 10M for wireless network card
	Display resolution	It can adapt to different resolutions such as 1024 * 768/1600 * 1280
Shared resource software running environment	operating system	WindowsXP/Win7/Win8
	Develop software requirements	1. Multimedia player software such as Windows Media Player, Flash Player 11.0 and JRE is provided 2. Provide JavaScript, IE Tester and other page development software environments 3. UI based design 4. Shared resource development realizes interactive application development of Flash AS3.0 5. Use Premiere format factory software for video editing and Camtasia 8.0 software for screen recording 6. CSS + DIV to beautify the interface, Flash animation to design two-dimensional images and so on

Basic data is resident data for the terminal application education and teaching management platform, and its huge data volume is not suitable for the communication mode of frequent interaction between the education and teaching management platform and the service portal. Therefore, the development of curriculum sharing resources needs to provide terminal object oriented basic data applications to obtain all basic data information visible under the user's authority at one time, And cache these data on the terminal

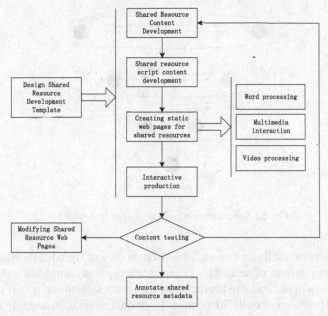

Fig. 4. Implementation process of online teaching shared content development

equipment, so as to reduce the frequent communication and interaction of large amounts of data during the operation of shared applications, improve the education and teaching management platform and the sharing terminal, use the response speed of the system, save the network traffic of terminal equipment, and effectively control and supervise the traffic and sharing process.

After the shared resource module is developed and provided to the database network platform, the shared resource conflict resolution method needs to be tested and analyzed. This paper uses the V-type test model, as shown in Fig. 5, and combines the actual situation of the software of the college education and teaching shared network platform to respond to the quality of the most real software system.

According to the schematic diagram of V-shaped test model in Fig. 5, the conflict resolution method test was carried out, and the page content check, link relationship check and E-browser test were carried out for the production effect of each courseware in a timely manner. This software can test the browser compatibility of courseware. The education and teaching sharing network platform performs the following test phases:

1. Unit test.

The education and teaching sharing network of secondary vocational schools belongs to the category of medium and large software according to the scale of software, so unit testing is very necessary in the process of software system development, and will become the basis for the normal development of the subsequent testing phase. The unit test of the platform should be completed by the developers of all R&D teams, which belongs to the function level test verification link, so that the developers can ensure the reliability of the functions themselves.

2. Function test.

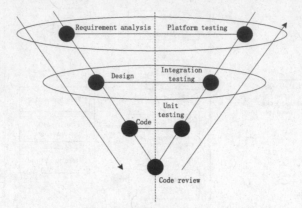

Fig. 5. Schematic diagram of V-type test model

Function test is a test link when each function module of the education and teaching sharing network platform of secondary vocational schools is completed independently. The function test will be conducted based on the unit test results to verify the accuracy and integrity of the functions of each independent functional module. In addition to verifying the quality of functional modules, the design of the software system also achieved the verification effect in the functional test phase. Functional testing is mainly performed and completed by testers.

3. Integration test.

Integration test is a joint test between modules implemented after all functional modules of the education and teaching sharing network platform of secondary vocational schools have been developed. The integration test is performed by testers on the basis of functional test. Through the integration test, it can verify the collaboration ability between the functional modules of the education and teaching sharing network platform, and at the same time verify the effectiveness of the software system design of the system.

4. System test.

The system test is a black box test conducted after the integration test of the education and teaching sharing network platform is completed. It is a process of further comparing and verifying the functions and requirements analysis results of the software system from a macro perspective. The system test is carried out by the tester, which aims to find the contradiction between the developed system functions and the requirements analysis results in a timely manner, and correct it in a timely manner, so as to improve the reliability of the system operation.

3.2 Result Analysis

Test method and process: unit test (white box) - integration test (black white combination) - system test (black box) - acceptance test accompany the software development process.

Test content: performance and pressure test. Test purpose: to test the stability, reaction speed, fault tolerance, compression resistance, etc. of the system. Test method: consciously give wrong or substandard input and test the output results of the system.Test approach: choose the black box test method, and randomly find teachers and students who do not understand the online teaching sharing resource platform to experience. The test results are shown in Table 2.

Table 2. Results of online teaching shared resource conflict resolution performance test

Operation	Expected results	Actual results	Conclusion
Log in to the system, and enter an inconsistent user name and an incorrect password	Give corresponding error prompts, such as your password is wrong, you have entered invalid characters, etc	Consistent with expected results	Normal
High frequency click the mouse at any position on the interface	The system will not be stuck. If you click to the function area, there will be a corresponding prompt, such as you have not selected to download files	Consistent with expected results	Normal
Halfway through the sharing operation, exit the system by force	You can log in again smoothly, and there will be no data that failed the last operation	Consistent with expected results	Normal
Share many large files at one time, and check all users of the platform	Normal sharing, expected to take a little longer	Consistent with expected results	Normal

According to the performance test results in Table 2, the performance test results of the online teaching shared resource conflict resolution method proposed in this paper are consistent with the expected results, meeting the teaching needs. On this basis, the error rate of shared resource conflict resolution in online teaching of design professional artworks is selected as the evaluation index for this experimental analysis, and the conflict resolution method of shared resource in online teaching of design professional artworks based on feedback integration proposed in this paper is set as the experimental group, and the traditional conflict resolution method is set as the control group. After the application of the two conflict resolution methods, the error rate of conflict resolution of online teaching shared resources for art design majors in X colleges and universities was measured and compared, and a comparison chart of evaluation indicators was drawn as shown in Fig. 6.

From the comparison results of the evaluation indicators in Fig. 6, we can see that after the application of the two methods, there are some differences in the corresponding online teaching shared resource conflict resolution effects. Among them, after the application of the shared resource conflict resolution method for online teaching of design professional artworks based on feedback integration proposed in this paper, the error rate of shared resource conflict resolution is significantly lower than that of traditional methods, with the maximum not exceeding 0.2%. From the comparison results, it is

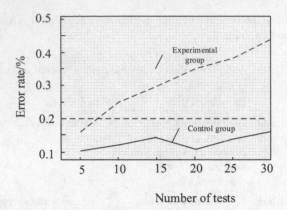

Fig. 6. Comparison Diagram of Evaluation Indicators

easy to see that the method proposed in this paper is highly feasible, effectively avoiding the conflict, asymmetry and other problems of online teaching resources of design professional artworks, and has significant advantages in application effect.

Select a group of design major students as the experimental subjects and provide them with a series of art online teaching and sharing resources, including design cases, video tutorials, learning materials, etc. Design conflict scenarios where multiple students simultaneously choose the same artwork for learning and creation, resulting in resource competition. Calculate the resource utilization rate, as shown in Fig. 7.

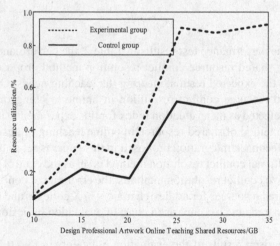

Fig. 7. Comparison of Resource Utilization Results

From Fig. 7, it can be seen that there are certain differences in resource utilization rates between the two methods after application. Among them, the resource utilization rate of the shared resource conflict resolution method for online teaching of design professional artworks based on feedback integration proposed in this article is significantly higher than that of traditional methods. From this comparison, it can be seen that the method proposed in this article has high feasibility.

4 Conclusion

Facing the limitation of educational resources, in order to improve the full use of existing resources and realize the optimal allocation of educational resources, colleges and universities practice sharing educational resources, which is the path choice for the healthy development of China's higher education system. This paper is based on a feedback integration method for resolving shared resource conflicts in online teaching of design related professional artworks. It combines feedback mechanisms and integrated design ideas to address potential resource conflicts that may arise during the online teaching process. By establishing a conflict resolution strategy model for online teaching shared resources and designing a collaborative service control protocol for teaching resource sharing, collaborative sharing and control of teaching resources have been achieved. At the same time, a shared resource integration and collaboration platform was designed based on feedback integration, providing a collaborative and shared environment for teachers and students. In the future, further research and improvement can be conducted on the conflict resolution strategy model of online teaching shared resources, utilizing more advanced algorithms and technologies to improve the efficiency and accuracy of conflict resolution.

Acknowledgement. Guangxi Education Science "14th Five-Year Plan" 2022 annual private higher education special project "Research on the integration path of art teaching resources for design majors in private universities: appreciation, storage, exhibition and marketing", Project number 2022ZJY3223.

References

1. Zabolotna, O., Zagoruiko, L., Panchenko, I., et al.: Teaching English vocabulary online: is the screen a barrier?. Adv. Educ. **2021**(17), 57–64 (2021)
2. Yalagi, P.S., Dixit, R.K., Nirgude, M.A.: Effective use of online teaching-learning platform and MOOC for virtual learning. J. Phys. Conf. Ser. **1854**(1), 012019 (8pp) (2021)
3. Liu, L., Tsai, S.B.: Intelligent recognition and teaching of English fuzzy texts based on fuzzy computing and big data. Wirel. Commun. Mob. Comput. **2021**(1), 1-10 (2021)
4. Lu, J., Gao, H.: Online teaching wireless video stream resource dynamic allocation method considering node ability. Sci. Program. **2022**, 1–8 (2022)
5. He, Y.: Design of online and offline integration teaching system for body sense dance based on cloud computing. J. Interconnection Netw. **22**(Supp05) (2022)
6. Tucker, B.V., Kelley, M.C., Redmon, C.: A place to share teaching resources: speech and language resource bank. J. Acoust. Soc. Am. **149**(4), A147–A147 (2021)
7. Zheng, H.Y., Ran, X.C.: Application of QR code online testing technology in nursing teaching in colleges and Universities. Sci. Program. **2021**(Pt.13) (2021)
8. Dong, X., Chen, X.: Research on online teaching of college teachers under the background of education informatization. MATEC Web Conf. **336**, 05005 (2021)

9. Eberle, J., Hobrecht, J.: The lonely struggle with autonomy: a case study of first-year university students' experiences during emergency online teaching. Comput. Hum. Behav. **121**(3), 106804 (2021)
10. Yang, Y., Zhang, M., Wu, B., et al.: Design of simulation system for discontinuity network in rock mass and its application in teaching. Comput. Simul. **39**(9), 5 (2022)

Design of Substation Battery Remote Monitoring System Based on LoRa Technology

Chen Zhao[✉], Dong Yang, Xiao Xu, Gongying Zhang, and Qirui Xu

Suzhou Power Supply Company, Suzhou 23400, China
17755711383@163.com

Abstract. The function module of the current substation battery remote monitoring system is generally one-way, and the monitoring range is limited, leading to the extension of the monitoring response time. Therefore, the design of the substation battery remote monitoring system based on LoRa technology is proposed. According to the actual monitoring needs and standards, first build the main controller MCU, access the A/D conversion circuit, establish GPRS monitoring sensor device, and complete the color design of the system hardware; On this basis, the multi-target form is adopted to break the restriction of the monitoring range, and the multi-target remote monitoring function module is established. The associated monitoring range is limited. At the same time, the monitoring storage database is constructed to complete the design of the system software. The final test results show that the final monitoring response time is better controlled below 0.5s for the operation of six batteries, indicating that this method has better monitoring effect, stronger pertinence and practical application value.

Keywords: LoRa Technology · Substation Battery · Remote Monitoring System

1 Introduction

The substation battery is often used as the backup power supply in the power supply system. It is a DC power supply. When the main power supply system fails, the secondary system in the power system maintains its normal work through the power supply of the battery. Therefore, the stability of the substation battery in the actual operation process and its discharged capacity in the discharge process are of great significance to ensure the normal operation of power equipment [1]. At present, the management and remote monitoring of the battery in the substation are mainly one-way. While collecting the daily system operation data set information, it is also necessary to test the total capacity of the battery through the discharge test. In the process, the battery is replaced in a periodic manner to prevent the power supply system failure caused by insufficient battery capacity, and further expand the actual monitoring range. The greater the coverage of one-way monitoring, the longer the time will be spent, and the periodic replacement of battery will result in waste of battery resources and economic losses [2]. Moreover, the traditional substation battery remote monitoring system itself lacks pertinence. In the

L. Yun et al. (Eds.): ADHIP 2023, LNICST 550, pp. 53–66, 2024.
https://doi.org/10.1007/978-3-031-50552-2_4

face of different power environments, the control effect of the monitoring area is also different. In order to ensure the change of the daily monitoring needs and standards of the substation, the monitoring program of the system is improved and optimized [3]. The so-called LoRa technology mainly refers to a low-power LAN wireless standard. Its name "LoRa" is Long Range Radio. The biggest feature is that under the same power consumption conditions, it spreads farther than other wireless methods, realizing the unification of low-power consumption and distance. Under the same power consumption, it extends 3–5 times the distance of traditional wireless radio communication [4]. Integrating this technology with the substation battery remote monitoring system can, to a certain extent, further expand the actual monitoring range, gradually build a flexible and changeable monitoring structure, and improve the quality and efficiency of monitoring [5].

In fact, in recent years, with the advance of smart grid, more and more substations have realized unattended operation, and the safe operation of substations is particularly important for the power system. After the evacuation of attendants, the demand for safety monitoring of batteries becomes very strong [6]. As a DC system that undertakes the tasks of secondary load power supply such as control, protection circuit and automatic device, its reliability will directly affect the normal operation of the substation [7]. At this time, the battery becomes the last line of defense of the DC system, which plays a particularly important role in the event of an accident in the substation. Using LoRa technology and intelligent information technology, the remote monitoring structure built can mark the abnormal position faster and more timely, collect the corresponding data and information, and transmit them to the preset position [8] through the channel. In addition, with the help and support of LoRa technology, the system can also monitor the internal resistance, temperature, voltage parameters, etc. of the substation battery in real time, help relevant personnel to analyze the performance of the battery, and grasp the operating status of the battery at any time [9]. In case of battery failure, the system will take alarm prompt measures to remind the maintenance management personnel to deal with the failure event in time to avoid accidents caused by battery failure, ensure the reliability and safety of the power supply system, reduce the possibility of battery failure caused by damage and aging of individual battery, and extend the service life of the battery. The application of LoRa technology has also enhanced the use of battery online monitoring system by maintenance personnel to a certain extent, avoided frequent on-site inspection, greatly reduced the workload, reduced injuries and accidents to relevant personnel, and laid a foundation for the subsequent development and innovation of the power industry and related technologies [10]. Therefore, the design and verification research of remote monitoring system for substation battery based on LoRa technology is proposed. Based on the actual remote monitoring requirements of the substation battery, a main controller MCU is constructed and connected to an A/D conversion circuit to reduce power consumption and cost. A GPRS monitoring sensing device is established and hardware design is completed. By utilizing LoRa technology, designing a multi-objective remote monitoring function module and an anomaly monitoring warning program, it is convenient to directly grasp the application status of substation batteries through the system. At the same time, a monitoring and storage database is constructed to facilitate the analysis of monitoring data, thus achieving the design of a substation battery remote monitoring system based on LoRa technology.

2 Design Hardware of Remote LoRa Monitoring System for Storage Battery

2.1 Main Controller MCU Design

The so-called MCU mainly refers to a kind of microcontroller unit (MCU), also known as single chip microcomputer or single chip microcomputer, which is to properly reduce the frequency and specification of the central process unit (CPU), and to properly reduce the memory, counter, USB, A/D conversion, UART, PLC, DMA and other peripheral interfaces, A hardware structure that forms overlay control.The selection of main controller MCU is also very important and critical for the design of monitoring system.It is possible to connect a single chip microcomputer into the hardware control structure, but it should be noted that the following aspects should be considered when selecting a single chip microcomputer: the reading and writing effect of the GPRS communication module SIM900, the control range of the LCD, and the correlation degree of the battery collection module. The single chip microcomputer that meets the above three items can be connected.Set a stable interface on the periphery of the MCU, select the SPI interface for GPRS communication, and the RS232 interface for communication with the S-BUS converter. Add a master control chip in the internal control structure to meet the requirements of system scalability as far as possible. On the basis of retaining the existing hardware structure, expand the new peripherals.The main controller selects the STC15F2K60S2 single chip microcomputer, sets the number of on-chip integration to more than 100000 times, and associates the Flash program memory with the RS232 interface communicated by the S-BUS converter to form a stable control program.Set basic control indicators and parameters, as shown in Table 1.

Table 1. Main controller MCU index and parameter setting table

Main controller MCU indicators	Controllable parameter standards	Measured parameter standards
Working current/mA	35	45
Voltage/V	220	220
Number of RS232 serial interfaces/piece	12	18
Low level/dBm	80	100
High level/dBm	220	260

According to Table 1, complete the setting and research of MCU indicators and parameters of the main controller.Next, the GPRS wireless communication module is set based on the actual monitoring requirements and standards. This part can be controlled by combining the operation status analysis of the substation battery.To set the working voltage range, it is generally required to control between 2.4 V and 5.5 V.The watchdog and timer protection devices are set in the hardware structure to gradually improve the anti-interference capability of the system. A reliable reset circuit is set inside the control circuit, and the program structure is encrypted before transmission to form a stable control hardware control system.

Then, based on this, a single chip computer of STC15F2K60S2 model is connected to the current hardware structure, and the initial SIM300 wireless module, alarm module, keyboard input module, RS232 to TTL level circuit, and LCD1602 with backlight are connected.A small control circuit is designated in the control circuit, which is composed of an external crystal oscillation circuit and a reset circuit.The oscillation frequency of the external quartz crystal is 22M, and the reset circuit selects the resistance capacitance reset circuit. Next, based on the actual hardware control needs, set the external control circuit of the main controller MCU, and increase the converter interface to the non-standard nine pin RS232 serial interface. The electrical level standard of the RS232 serial interface is inconsistent with the TTL level of the MCU, which is easy to cause the current and voltage loops to press, To reduce the operation capacity of the battery, in order to achieve the communication balance between the main controller MCU and the converter, a RS232 TTL level access circuit is designed. The MAX232 chip is used to complete the setting and guidance of the serial interface circuit, and the TTL level 0V or 5V is converted to 3V to 15V or −3V to −15V through MAX232 to form a multi-directional controllable main controller MCU communication unit structure, as shown in Fig. 1.

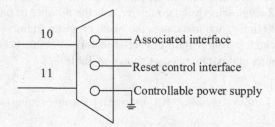

Fig. 1. Structure diagram of main controller MCU communication unit

According to Fig. 1, complete the design and research of the main controller MCU communication unit structure. Next, adjust the working voltage at this time to 4.5V–5.5V, and the working voltage at 5V is the optimal mode. When the working voltage is 5V, the working current is 2.0 mA. Set RS as data and command selection, R/W as read/write selection, and E as enable signal. When RS of battery is low level and R/W is high level, LCD of main controller MCU is in reading state; When RS is low level, R/W is low level, and enable signal E is high pulse. At this time, the main controller will write the instruction code into 1602 and convert it to the data reading state, strengthen the application effect of the main controller MCU, and complete the basic design.

2.2 A/D Conversion Circuit Design

After completing the design of the main controller MCU, the next step is to build the A/D conversion circuit. The remote monitoring of the storage battery generally requires the cooperation of the sensor, which adopts the form of SBUS bus redundant connection to ensure that after the sensor collects the working parameters of the storage battery, it obtains the SBUS data signal, realizes the communication with the main controller, sends the data to the main controller, and realizes the action of SBUS to RS232 circuit. The internal control circuit needs to use ICL3221, the main chip of ICL3221, and set RS232 interface. In this part, it should be noted that in order to ensure the application control effect of RS232 interface, one driver and two receivers can be connected to the CL3221 and ICL3221 main chips. The full duplex communication mode is used to set the basic A/D conversion circuit form and calculate the transmission rate of the circuit, as shown in Formula 1 below:

$$H = \sum \eta i + \kappa^2 \Re c \tag{1}$$

In Formula 1: η indicates the communication control range, i indicates the control times, κ indicates the serial port protocol distance, $\Re$ represents the lowest voltage, c indicates the sensing coverage. According to the above measurement, the calculation of the circuit transmission rate is completed. Adjust the transmission rate standard of the battery remote monitoring system to control it between 120 ~ 250kbps, and the working voltage of the modified times is 3 V~5.5 V, so as to ensure that the working circuit of ICL3221 is in normal and stable operation state under the condition of low power consumption. At this time, set the serial port of RS232 communication protocol between ICL3221 and MCU, set the converter in the A/D conversion circuit, process the collected data signal, and send the serial port protocol through the S-BUS bus. Under normal conditions, SBUS signal enters from the SBUS TXD of the circuit, and is sent to ICL3221 from the T1-IN terminal after being amplified by the connected triode OP07AJ and processed by the circuit, ICL3221 receives the data conforming to the RS232 communication protocol standard. At this time, the A/D conversion circuit can control the transmitted data at the same time, collect analog signals, read the data using the main controller, and achieve specific control and monitoring functions through an A/D conversion circuit.

Then, based on this, an analog resistor is connected in series to the output circuit of the current sensor to convert the current signal into a voltage signal. AD7705 is selected as the main chip of AD conversion, which has low power consumption, low cost, accuracy and resolution to meet the system requirements. At this time, the control value of A/D conversion circuit is set according to the actual remote monitoring requirements and standard changes, as shown in Table 2.

Set and analyze the control value of A/D conversion circuit according to Table 2. At this point, the corresponding TCP/IP protocol stack can be established through the internal integration of SIM300, and the control instructions of TCP/IPAT circuit can be expanded by integrating the actual requirements and standard changes, and the AT instruction set can be used to conduct differentiated control processing on the conversion circuit, so as to better realize the transmission and application of GPRS data, further

Table 2. A/D conversion circuit control value setting table

Conversion circuit indicators	Fixed standards	Controllable standard
Main chip type	SIM300	Integration SIM300
Conversion rate	1.3	1.5
Control frequency/time	12	16
Minimum voltage/V	5.5	6.5
Circuit expansion times/time	8	12
Number of pins/piece	32	64
Working voltage/V	4	4.5

improve the design of hardware structure, and reduce the occurrence of error control, control delay and other problems.

2.3 Design of GPRS Monitoring Sensor Device

After completing the design of A/D conversion circuit, the next step is to build and access the GPRS monitoring sensor device.Generally, before the design of GPRS monitoring sensor device, it is necessary to define the range and distance of sensor coverage, and ensure the reliability and stability of the circuit when making PCB board.Several sensing layers are designed and the sensing device is designed in four stages, as shown in Table 3.

Table 3. Setting table of multi-stage sensing device

Sensing stage	Number of TC15F2K60S2 chip pins/piece	Peripheral expansion control ratio
Phase 1	12	1.13
Phase 2	12	1.25
Phase 3	16	1.27
Phase 4	18	1.31

Set the multi-stage sensing device according to Table 3. Then, based on this, the PCB circuit board was designed. However, in the actual design process, attention should be paid to the crystal oscillator operation state of the MCU. At the same time, a sensor monitoring device should be set close to the STC15F2K60S2 chip pin for this purpose to minimize its interference to the surrounding devices. Try to use SMD components, ground wires, and power wires to walk together. The ground wire, power wire, and signal wire should not form a loop. The circuit part that needs external expansion module should be placed on the periphery of the circuit board to facilitate wiring with the expansion module, so that the application coverage of GPRS monitoring sensor device in the system main controller PCB hardware structure can be improved. Next, according to the actual needs and standards, set the decoding of the sensor return data frame, as shown in Table 4.

Table 4. Decoding setting table of returned data frame of sensor device

Directional parameters	Data A + Data B	Exponential, mantissa	Price
Voltage	0 x 55, 0 x 00	0.1010,101	13.12
Voltage	0 x 42, 0 x 01	0.1011,102	14.61
Temperature	0 x 66, 0 x 02	0.1012,103	15.02
Impedance	0 x 81, 0 x 03	0.1013,104	15.37

According to Table 4, complete the setting of decoding the returned data frame of the sensing device. At this time, analyze the operation of the battery. Under the same conditions, the greater the discharge current of the battery, the greater the current is due to the existence of its own internal resistance, the more power it consumes, and the less power it can supply for the external circuit. Therefore, we can take GPRS monitoring sensing correction measures to a certain extent for the residual capacity of the battery when discharging with different discharge currents. The residual capacity of the battery will be affected by temperature, so first correct the temperature and calculate the corresponding corrected capacity, as shown in Formula 2 below:

$$K_{corr} = \frac{t_{act}^2}{l(t_{init} + t_{std})} \tag{2}$$

In Formula 2: l indicates the number of sensing times, t_{init} and t_{std} represent the current correction deviation and the actual correction deviation respectively, t_{act} indicates the sensing orientation value. According to the above measurement, the calculation of the correction capacity is completed. According to the obtained values, the control and monitoring indicators of GPRS monitoring sensors are adjusted to form a stable remote monitoring structure, and the design of the system hardware is completed.

3 Design of Remote LoRa Monitoring System Software for Storage Battery

3.1 Multi Target Remote Monitoring Function Module Design

After completing the design of the system hardware, the next step is to integrate the LoRa technology to design the software in the battery remote monitoring system. Firstly, the function module of multi-target remote monitoring is designed. Generally, in order to increase the controllability of the system and improve the monitoring efficiency and quality, the system is divided into monitoring modules such as system management, basic data management, battery condition monitoring, battery operation monitoring, battery charge and discharge monitoring and analysis, battery internal resistance analysis, battery performance analysis, status alarm, and comprehensive report. Through the organic combination of these functional modules, the online monitoring and analysis of the battery can be realized after the effectiveness verification and scientific analysis of the real-time monitoring data. The specific structure is shown in Fig. 2.

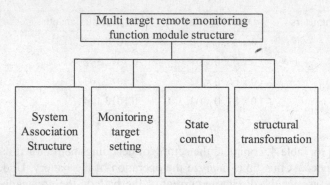

Fig. 2. Structure diagram of multi-target remote monitoring function module

According to Fig. 2, complete the design and analysis of the structure of the multi-target remote monitoring function module. Next, integrate the LoRa technology to analyze and verify the key application function modules. First, the basic system management is designed and analyzed. This part mainly manages system users, permissions and other information. The setting content mainly includes the sub modules of menu management, role management, company management and user management.

The system administrator can assign corresponding roles to each user by adding users on a daily basis to control the user's permissions. At the same time, the system administrator can set the initial control indicators and parameters according to the coverage of different function monitoring modules, as shown in Table 5.

According to Table 5, set and study the index parameters of the multi-target remote monitoring function module.Next, the basic data management module is designed. This part usually needs to be divided into multiple layers, mainly providing some basic data for system operation. Generally, the number of daily reaction battery operation and some frequently used data will be uniformly maintained. In this system, the basic data management function mainly includes two sub modules, namely, parameter setting module and battery pack maintenance module. Basic data management mainly includes sub function structure and directional function structure. All of them play a targeted role in the daily operation and maintenance of the battery.

Then, based on this, the main monitoring function module of battery operation is designed. The battery has strict requirements for operation, and serious consequences will be caused if there are environmental problems, so it is very important to monitor its operating parameters. The traditional battery operation monitoring is mainly completed manually, and the status of the battery is measured regularly. With the development of technology, using equipment to monitor the battery system in real time and online will gradually replace manual testing as the main means of battery testing. However, most of the automatic monitoring is only limited to the monitoring of battery voltage,

Table 5. Index Parameters of Multi target Remote Monitoring Function Module

Remote monitoring function module	Define	Controllable value
USER	Daily control and setting of users	User management and permission control
Permission settings	Setting of system control permissions	Application of system functions
Roles	User + administrator	Administrator + User Settings
menu management	Basic functional control	Fixed application function settings
Units Manager	Monitoring Content Management	Internal control structure setting
Sub module design	Related module design	Sub module control

current, temperature and other operating parameters, which is difficult to achieve real-time monitoring of battery performance. Remote monitoring connects the monitoring equipment with the battery, collects and reports the battery voltage, charge and discharge current, temperature or individual voltage and other operating parameters in real time, and conducts charge and discharge management, effectively making up for the weakness of manual inspection. Multi objective remote monitoring module can further expand the actual monitoring range, and form more controllable monitoring standards. The control of monitoring error shall be strengthened to the greatest extent to strengthen the application level of each functional module.

3.2 Abnormal Monitoring and Warning Program Design

After the design of the multi-target remote monitoring function module is completed, the next step is to build an exception warning program combined with LoRa technology.In the daily operation process, the system will combine the actual situation, provide abnormal conditions for the system operation, and give real-time alarm. When a battery is found to have abnormal conditions, the system will automatically give an audible alarm. The thresholds of these alarms can also be modified by the user. Next, on this basis, in combination with the abnormal state, we use professional software to carry out abnormal statistical processing, and summarize and integrate the corresponding data and information. The system can make classification based on the abnormal monitoring conditions and generate corresponding statistical reports and test reports in combination with LoRa technology. It can generate corresponding battery operation reports by month, month and day. Through these reports, the performance and operation of on-site batteries can be analyzed, providing relevant data support for further decision-making.

The data analysis and statistical process of the statistical report are not fixed, but make targeted adjustments with the change of the actual situation, with strong pertinence and stability. Not only that, combined with LoRa technology, the system will also collect relevant information about the battery in the front-end substation, and then judge the collected data. If there is an exception, there will be an exception information table. If it is normal, there will be an exception information table in the history information table. The system will automatically generate the corresponding operation data table according to the time set by the user in combination with LoRa technology.

3.3 Monitoring Storage Database Design

After completing the design of the exception warning program, the next step is to build the monitoring storage database. Monitoring database design is a series of processes that store records previously stored manually in a database, analyze and extract relevant records, and then design according to database requirements in combination with LoRa technology. When designing the database, you should first design the table according to the storage requirements, then design the fields and their storage types, as well as the primary key, foreign key, non empty constraints and other relationships. Combined with LoRa technology, LoRa multi-dimensional recognition sensing program is constructed to correlate with the database. The design process is also quite complex. Generally, people with database design experience design the database according to the functional design of the project. In the process of design, the following issues also need to be considered, specifically as follows: First, consider the scalability of database tables, comply with the three paradigm principles of data design, try to ensure that a table only stores data of one entity, and reduce data coupling. The second is to consider the integrity of the database. When designing, try to consider the usefulness of the data in each field to reduce the redundancy of data. The third is the change of database performance. Consider whether to establish indexes separate meters, etc., so as to improve the running speed. For database design, it is very important to ensure data integrity and reduce data redundancy by combining LoRa technology. It is very important to design a reasonable database table structure. Generally, the database design needs to be reviewed after completion. The system database designed this time combines LoRa technology, uses the MYSQL database association form, and the database table tool designed is Power Designer as shown in Table 6.

Table 6. Database Table Tool Information Table

Database Field Description	Types of	Notes
ID	One-way	Address
ORG-NAME	Attributive name	Name
ORG-TEL	Bidirectional	Phone
ORG-ADDRESS	Int	Work address
NOTE	Other	Memo

According to Table 6, complete the setting and analysis of database table tool information. Next, combine LoRa technology to build a one-way control program in the monitoring database to form a more controllable monitoring structure. On this basis, the personnel information table needs to be used as the system user information storage, and personnel information is the primary prerequisite for system operation. The personnel information controlled in this part generally shows the personnel information of the substation. The users of the system are roughly divided into system administrators and operation and maintenance personnel. The system administrator is mainly responsible for managing all functions in the system. It can operate all functions in the system and view all monitoring information; Maintenance personnel can see the functions of battery performance analysis, status monitoring, statistical report, status alarm, etc.

The personnel information table set in the monitoring repository mainly maintains information such as the name, real name, password, contact number, contact address, affiliated unit, registration date, and other relevant information used when logging in. When designing the database, we also designed the field length and data type of the personnel information table, which is mainly used to maintain the values that are often used in the system and will change frequently, so that users can flexibly set the parameter values according to the use conditions, thus reflecting the flexibility activity. Reduce the need to modify the code or database when a value changes. Users can directly master the application of the substation battery through the system, which is more convenient. In addition, the monitoring database stores information data, which is also convenient for subsequent monitoring and analysis.

4 System Test

This time is mainly about the analysis and design of the actual effect of the remote monitoring system for the storage battery in the substation based on LoRa technology. Considering the authenticity and reliability of the final test results, the analysis is carried out by means of comparison. Substation G is selected as the main target object for the test, and professional equipment and devices are used to collect relevant system design data and information. According to the actual design requirements and standards, the final system test results are analyzed. Next, the basic test environment is built based on LoRa technology.

4.1 Test Preparation

In combination with LoRa technology, this time is mainly to build the test environment for the battery remote monitoring system of G substation. First, you need to set the background data management server (DMS). The service device is mainly the core of the battery online monitoring system, which is responsible for monitoring the operation of the whole system. Its functions include: system initialization, setting BDCU working status, reading monitored data, graphic display, database processing, data analysis and other modules. First, initialize the system clock, set internal environment variables and issue initialization instructions to BDCU in combination with the designed battery monitoring requirements and standards. On this basis, it is also necessary to set BDCU working status: start or stop BDCU operation. And build association with the program

Table 7. Battery Operation Monitoring Reading Table

Describe	Offset start address	Offset End Address
Number of batteries/piece	12	16
Battery monitoring voltage average value/V	12	16
Average value of battery monitoring resistance/Ω	3.2	4.1
Internal resistance of battery/Ω	5.5	6.2
Monitoring time/s	1.01	0.82
Directional monitoring difference	2.72	3.62
Actual monitoring difference	1.92	2.71

reading monitoring data to read the battery operation data monitored by BDCU, as shown in Table 7.

Complete the setting of the battery operation monitoring reading program according to Table 7.Next, set the graphic display structure to display the status of each battery cell in a graphic manner. Audible and visual alarms will be given to battery cells that have detected excessive data, and the monitored values include internal resistance, voltage, etc. Then, the acquired data and information will be transferred to the database for equivalent processing. Store historical data, provide original data for data analysis, analyze the trend of historical curve of internal resistance of specific battery cells according to the historical database, arrange battery cells in the order required by specific requirements, and output backward batteries in the report. The data server (DMS) is connected to the internal network of the substation to provide database support to authorized network users, so that these users can get the battery data monitored by the system in real time for subsequent use. So far, the basic test environment has been set. Next, specific test research is carried out in combination with LoRa technology.

4.2 Test Process and Result Analysis

After completing the establishment of the basic test environment, next, combined with LoRa technology, the remote monitoring system of G substation battery is measured and analyzed. First, six batteries are selected as the main target objects for testing in the substation, and a corresponding number of monitoring nodes are set around the batteries. The nodes are connected and overlapped with each other to form a cyclic monitoring structure. The acquired data is processed and transferred to the display unit program for subsequent use.

Through in-depth data analysis of the voltage, current, internal resistance, temperature, etc. of the substation, the system makes reasonable adjustment and correction to ensure that the battery is in the most stable state, comprehensively judges the condition of the battery and reasonably and effectively predicts its performance. The measurement is carried out in three stages. Three distances of 14 km, 20 km and 24 km are set to test the remote monitoring system, and the monitoring response time is calculated, as shown in Formula 3 below:

$$T = d^2 - \left(\vartheta + \sum dw\right) - \varphi \tag{3}$$

In Formula 3: d indicates the directional monitoring range, ϑ indicates the overlapping monitoring range, w indicates the monitoring times, φ indicates directional deviation. According to the above determination, complete the analysis of the test results, as shown in Table 8.

Table 8. Comparison and Analysis of Test Results

Measuring the battery	14 km monitoring response time/s	20 km monitoring response time/s	24 km monitoring response time/s
Storage battery1	0.21	0.37	0.41
Storage battery2	0.25	0.31	0.47
Storage battery3	0.23	0.36	0.42
Storage battery4	0.26	0.37	0.42
Storage battery5	0.29	0.32	0.47
Storage battery6	0.26	0.37	0.41

According to Table 8, the analysis of the test results is completed: for the operation of six batteries, the final monitoring response time is better controlled below 0.5s, indicating that the monitoring effect of this method is better, more targeted, and has practical application value.

5 Conclusion

The above is the design and verification research of the remote monitoring system of the substation battery based on LoRa technology. In the hardware part of the substation battery remote monitoring system, the communication unit structure of the main controller MCU is designed, and an A/D conversion circuit is constructed based on actual monitoring needs to reduce system errors and delays. Select a GPRS monitoring sensing device to form a stable remote monitoring structure. In the system software section, a multi-objective remote monitoring function module, abnormal monitoring warning program, and monitoring storage database were designed to improve the monitoring effect and achieve the monitoring function of the system. Through experiments, it has been verified that the monitoring effect of this system is better and more targeted.

Compared with the initial battery monitoring mode, this integrated LoRa technology has designed a more flexible, variable and stable detection structure. At the same time, combined with the actual monitoring needs, further expand the scope of monitoring, strengthen the monitoring efficiency, and provide reference and theoretical reference for the innovation and upgrading of subsequent related systems. The future substation battery remote monitoring system can further research intelligent monitoring and more intelligently monitor battery status to improve battery efficiency and lifespan.

References

1. Mellit, A., Benghanem, M., Herrak, O., et al.: Design of a novel remote monitoring system for smart greenhouses using the internet of things and deep convolutional neural networks. Energies **14**(16), 5045 (2021)
2. Li, Q., Zhang, X., Xie, S., et al.: Research on remote monitoring based on cloud for electric air compressor of rail vehicle. J. Phys. Conf. Ser. **1986**(1), 012063 (2021)
3. Kamruzzaman, M.M., Alanazi, S., Alruwaili, M., et al.: Fuzzy-assisted machine learning framework for the fog-computing system in remote healthcare monitoring. Measurement **195**, 111085 (2022)
4. Huang, X., Liu, Y., Zhou, J., et al.: Garment embedded sweat-activated batteries in wearable electronics for continuous sweat monitoring. NPJ Flexible Electron. **6**(1), 1–8 (2022)
5. Wang, J., Zhao, Q., Ye, Q.: Design of remote monitoring system for substation DC power supply under the background of big data. J. Phys. Conf. Ser. **2037**(1), 012005 (2021)
6. Ahmid, M., Kazar, O., Kahloul, L.: A secure and intelligent real-time health monitoring system for remote cardiac patients. Int. J. Med. Eng. Inform. **14**(2), 134–150 (2022)
7. Mieszek, M., Mateichyk, V., Tsiuman, M., et al.: Information system for remote monitoring the vehicle operational efficiency. IOP Conf. Ser. Mater. Sci. Eng. **1199**(1), 012081 (2021)
8. Perl, L., Meerkin, D., D'Amario, D., et al.: The V-LAP system for remote left atrial pressure monitoring of patients with heart failure remote left atrial pressure monitoring. J. Cardiac Fail. **28**(6), 963–972 (2022)
9. Cheung, C.C., Lee, B.K.: From improving survival to cost savings for the health system-remote monitoring for all? Can. J. Cardiol. **38**(6), 712–714 (2022)
10. Mei, J., Zhou, P., Liu, R., et al.: Design and simulation of an conical array with variable top angle for ultra-wide-band biology radar. Comput. Simul. **39**(12), 275–279 (2022)

Optimization Scheduling Algorithm of Logistics Distribution Vehicles Based on Internet of Vehicles Platform

Zhuorong Li[1(✉)], Yongchang Yao[2], Haibo Zhang[1], Nan Li[1], and Jianmei Sun[1]

[1] Dalian University of Science and Technology, Dalian 116000, China
13802016671@163.com
[2] Chongqing Vocational Institute of Tourism, Chongqing 409099, China

Abstract. In the phase of logistics distribution vehicle scheduling, the distribution time and travel distance of vehicles are relatively long due to the influence of the real-time change attribute characteristics of the actual traffic environment state. Therefore, this paper proposes an optimization scheduling algorithm for logistics distribution vehicles based on the Internet of Vehicles platform. The Internet of Vehicles platform including acquisition layer, transmission layer, data layer and application layer is constructed to achieve the acquisition and analysis of real-time information of the actual traffic environment status. In the logistics distribution vehicle scheduling phase, the objective function of comprehensively arranging the number of vehicles, the total distance traveled by the distribution vehicles and the time window constraint punishment is constructed. After the objective function is input into the Internet of Vehicles platform, the greedy algorithm strategy is used to achieve the optimal scheduling of vehicles. In the test results, the design algorithm achieves the goal of shortening the vehicle delivery time and driving distance without considering the results under traffic conditions and the scheduling effect under dynamic conditions. To sum up, the optimization scheduling algorithm of logistics distribution vehicles based on the Vehicle-to-everything platform can help optimize the travel distance and distribution time of vehicles, and improve the efficiency and accuracy of logistics distribution.

Keywords: Internet of Vehicles Platform · Logistics Distribution Vehicles · Optimal Scheduling · Number of Vehicles · Total Distance Traveled · Constraint Punishment

1 Introduction

At present, the advantage of logistics development is the increasing scale of the logistics industry, in which the proportion of the logistics industry in the gross domestic product is increasing, And drive the increasing number of logistics employees [1]; The logistics service capability has been significantly improved, and has been developing towards professional and social services: the technical conditions have been significantly improved [2], the establishment of information systems, cloud computing, Internet of Things and

L. Yun et al. (Eds.): ADHIP 2023, LNICST 550, pp. 67–83, 2024.
https://doi.org/10.1007/978-3-031-50552-2_5

other information technologies have been preliminarily applied; The infrastructure network is gradually improved, including railway transportation, port transportation, airport transportation, etc.; The logistics development environment has been continuously optimized, and it is proposed to vigorously develop the modern logistics industry [3]. At present, the main task is to improve the professional and social service level of logistics, develop third-party logistics, adopt modern management concepts, and improve logistics service capabilities; Strengthen the construction of logistics informatization [4], apply advanced information technology as soon as possible, establish logistics information system, and promote the development of logistics information sharing platform [5]. Establish an Internet of Vehicles platform to collect logistics resources, store and process logistics data, realize logistics information sharing [6], save resources and improve logistics service capability. The Internet of Vehicles platform is based on cloud computing and relies on the efficient processing capability provided by cloud computing technology to realize the virtual integration of physical resources. Through network technology, massive logistics information will be integrated as an information sharing platform for enterprises to use, and the resource pool formed by virtualization of basic resources will be uniformly scheduled and managed. The platform is used to operate logistics resources and customer resources, and it is processed through cloud computing technology to provide an optimized route for logistics distribution, improve the logistics transportation process, improve the transportation efficiency, reduce the cost of logistics enterprises, and improve the utilization rate of resources, which is of great significance to the development of the logistics industry [7]. An important prerequisite for the stable development of the modern logistics industry is to reduce the logistics transportation cost, and the key to solving the transportation cost problem of the logistics industry is to optimize the logistics system [8]. The key step of logistics system optimization is the reasonable scheduling of logistics distribution vehicles. By optimizing the scheduling of distribution vehicles, enterprises can reduce transportation costs, improve customer service levels and economic benefits, and thus obtain more profits. The logistics vehicle optimal scheduling problem is also known as the vehicle routing problem (VRP) [9]. This problem was first proposed by Dantzig and Ramser scholars in 1959. It focuses on providing goods distribution services to some customers with different quantities of goods demand by using a fleet (several vehicles) from the distribution center. Arrange an appropriate driving route for this fleet, comply with certain constraints, and complete distribution to meet customer needs, The arranged driving routes reach the goals such as the shortest total driving distance, the lowest total transportation cost or the shortest total service time [10]. Therefore, it can be described as that distribution vehicles start from one (or more) distribution centers and pass through several distribution points (randomly distributed) in a certain order under various constraints, so as to ensure that each distribution point is served by distribution vehicles and only one vehicle serves.

Combined with the above discussion, this paper proposes a logistics distribution vehicle optimal scheduling algorithm based on the Internet of Vehicles platform. By utilizing the collection, transmission, data, and application layers of the Internet of Things platform, real-time information on traffic environment status is obtained and analyzed to address the impact of actual traffic environment changes on vehicle delivery time and driving distance. This algorithm constructs an objective function that comprehensively considers the number of vehicles, total distance traveled, and time window constraints in the vehicle scheduling stage, and utilizes a greedy algorithm strategy to achieve optimal vehicle scheduling.

2 Optimization Algorithm Design of Logistics Distribution Vehicles

2.1 Structural Design of Internet of Vehicles Platform

At present, the demand for logistics is increasing day by day, and the cooperation between logistics enterprises is very little [11]. The reuse of human, material, financial and other resources is increasingly high, resulting in waste of resources. The government urgently hopes to promote the development of public logistics information service platform, encourage cooperation between enterprises, and achieve logistics information sharing. In the face of huge logistics resource storage and calculation, it is necessary to establish a uniformly managed Internet of Vehicles platform, which can gather logistics resources together, make more reasonable use of them, promote the better development of the logistics industry, and provide better services for other related industries, which has important social significance [12].

The Internet of Vehicles platform built in this paper is a public information platform. It collects information related to logistics activities for various industries, and improves the efficiency of logistics resource utilization through business processing, saves unnecessary resource reuse, saves logistics costs to a large extent, and promotes the development of logistics economy [13]. The logistics public information platform needs to have huge logistics information storage capacity and business logic processing capacity. It provides a perfect technical solution for the establishment of the platform by taking advantage of the characteristics of cloud computing, such as workflow standardization and intelligent decision-making [14]. The construction of the Internet of Vehicles platform requires computer network communication technology and database management technology to realize logistics information [15].

Share technology, equipment and other resources to achieve resource integration, including cloud computing, Internet of Things, GPS and other resources.

Advanced technology, automatic collection and storage of logistics information, through the application of computer information technology.

The function of the cloud platform is more powerful, and the logistics system is more perfect. The logistics distribution vehicle scheduling system based on the cloud platform studied in this paper uses the Internet of Things technology to collect logistics data, The collected data is stored and processed. The overall structure of the system can be divided into acquisition layer, transmission layer, data layer, application layer, etc. Four module representation.

Acquisition layer: mainly completes the conversion of data from the physical world to the information world. Logistics distribution vehicle scheduling system.

The data to be collected include cargo information, customer information, cargo logistics information, vehicle information, order information and other relevant information Interest.

Transport layer: to complete the transmission of logistics data, the main technical means are: broadband connection, WIFI, 5G mobile.

Communication technology, Internet of Things, etc., as shown in Fig. 1.Its purpose is to connect the acquisition layer and the data layer, and finally collect the data of the physical world to the cloud computing platform through the transport layer.

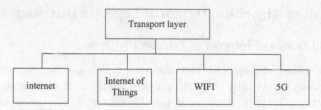

Fig. 1. Transport Layer Structure Diagram

Data layer: mainly used for the storage of logistics data, receiving data from the acquisition layer and storing it on the Internet of Vehicles platform.

According to the requirements of the application layer, the stored data is transferred to the application layer. Logistics data can be stored in SQLServer.

MySQL, Oracle and other databases, and SQLServer 2008 database is used for storage.

Application layer: also known as platform display layer, it is designed according to user requirements, and generally includes some information systems, including cargo information management system, vehicle information management system, logistics scheduling system, etc., as shown in Fig. 2. According to the query data specified by the user, query in the database through the service program, and feed back the results to the main user interface for presentation to the user.

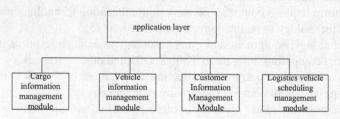

Fig. 2. Application Layer Settings

Under the above architecture design, the acquisition layer obtains corresponding data through sensors or other service platforms, and calls the interface in the data layer for data transmission. In the transmission process, the data can be transferred by registering the service with the data layer and completing the registration after obtaining the corresponding permissions. The data application is the opposite of the collection process. You need to build a computing instance first. After the instance is initialized, you can obtain the corresponding data from the cloud storage platform according to the application needs. You can directly analyze the obtained data, or create a database as needed to complete the data application.

In this way, the construction of the Internet of Vehicles platform is realized, and the acquisition of traffic vehicle information and the analysis of traffic operation status are realized, providing a reliable data basis for the subsequent logistics distribution vehicle scheduling.

2.2 Logistics Distribution Vehicle Scheduling Based on the Internet of Vehicles Platform

When carrying out logistics distribution tasks, the objective demand of customers is received. This paper introduces the time window mechanism, and on this basis, the distribution vehicle scheduling problem is split.

$$\sum_{j=1}^{N} x_{ijk} = \sum_{j=1}^{N} x_{ijk} \leq 1, i = 0, k \in \{1, 2, ..., K\} \tag{1}$$

$$\sum_{i=0}^{K} d_i \sum_{j=1}^{N} x_{ijk} = \leq q_k, k \in \{1, 2, ..., K\} \tag{2}$$

$$\sum_{i=0}^{K} \sum_{j=1}^{N} x_{ijk}(t_{ij} + s_i + w_i) \leq ET_0 \tag{3}$$

$$\sum_{i=0}^{K} \sum_{j=1}^{N} x_{ijk}(t_{ij} + s_i + w_i + t_i) = t_j, j \in \{1, 2, ..., N\} \tag{4}$$

$$ST_i \leq w_i + t_i \leq ET_i \tag{5}$$

Among them, t_i It means that the delivery vehicle drives to the customer i Time of, w_i It means that the delivery vehicle stops at the customer i Time of, K Indicates the total number of vehicles owned by the distribution center, N Indicates the total number of customers served by the distribution center, t_{ij} Represent customer i To customers j Vehicle travel time, d_i Represent customer i Demand, q_k Indicates the delivery vehicle k Maximum load of ST_i, ET_i Represent customer i's receiving time range, ST_i Is a customer i The earliest receiving time, ET_i Represent customer i Latest receiving time, s_i It means that the delivery vehicle is delivered to the customer i Time taken for unloading and delivery.

$$x_{ijk} = \begin{cases} 1 \\ 0 \end{cases} \tag{6}$$

where, when the vehicle k By customer i Drive to customer j When, then x_{ijk} value is 1, otherwise it is 0.

In the above formula, (1) represents the constraint condition of each vehicle starting from the distribution center and returning to the distribution center finally; (2) It means that the total volume or weight of goods to be delivered to customers for the loading of distribution vehicles cannot exceed the maximum loading capacity of distribution vehicles; (3) Is the constraint of the maximum travel time of distribution vehicles; (4) Indicates that the delivery vehicle has arrived at the customer j Time at is equal to the delivery vehicle arriving at the customer i Time at t_i, in the customer i Waiting time at w_i, in the customer i The sum of the service time, (5) represents the constraint of the customer's time window, and the delivery vehicle arrives at the customer i The time of plus the waiting time is within the customer's time window.

On this basis, this paper sets the weight value w_1, w_2, w_3 The objective function of this paper is composed of the total number of vehicles, the total distance traveled by vehicles and the penalty value, as shown in Formula (2–10).

$$\min(w_1 \sum_{i=0}^{K} \sum_{j=1}^{N} x_{ijk} + w_2 \sum_{i=0}^{K} \sum_{k=0}^{N} \sum_{j=1}^{N} c_{ij} x_{ijk} + w_3 p(t_i)) \tag{7}$$

Among them, c_{ij} Represent customer i To customers j Of vehicle travel, w_1 To arrange the weight of the total number of vehicles, w_2 Is the weight of the total distance traveled by distribution vehicles, w_3 It is the weight of punishment due to time window constraints, $p(t_i)$ represents the objective function of starting the number of vehicles, and the final optimization goal is to arrange the number of vehicles, the total distance of distribution vehicles, and the punishment due to time window constraints. The sum of the three weights is the smallest.

On this basis, input the objective function into the Internet of Vehicles platform built in Sect. 1.1. First, initialize the basic data, including the set population size, cross mutation probability, evolution algebra, maximum vehicle load, maximum vehicle travel distance, customer coordinates, customer demand, distance between customers, and other information. Initializing the vehicle information is to define a two-dimensional array vehicle [K, 3] with multiple rows and three columns. The row number represents the serial number of the vehicle. The first column represents the maximum loading capacity of the vehicle, the second column represents the maximum driving distance of the vehicle, and the third column represents the driving speed of the vehicle. Initializing customer demand information is to read the coordinates and demand information of customer points; Use a pair of one-dimensional arrays x [] and y [] to save the X axis coordinates and Y axis coordinates of each customer point, use the one-dimensional array guestDemand [] to save the demand of each customer point, use the distance formula between two points to calculate the distance between the distribution center and customers, and between customers, and save it in the two-dimensional array guest distance [,] for subsequent calculation, Use the paintPoint () method to draw the customer points on the graphical interface, and use the paintLine () method to draw the lines between customer points to visually display the line information.

Secondly, initGroup is the initial population. The general genetic algorithm obtains the initial population by randomization. The resulting population is not highly adaptive due to the randomness of individuals, and has a certain impact on the convergence speed and solution quality of the algorithm. In this paper, according to the characteristics of greedy algorithm that can quickly solve the local optimal solution, individuals with high fitness can be obtained by initializing the population. Adding greedy algorithm strategy can improve the fitness of the initial population to a certain extent, and accelerate the convergence speed of the algorithm. The specific implementation process of population initialization is to initialize with an empty two-dimensional array old Group. The number of rows represents the number of chromosomes, and the columns represent the gene fragments of chromosomes, that is, the customer's order. For each chromosome, first assign a random number to the first gene, which is a natural number from 1 to the length L of the chromosome. Then compare the next gene with the previous genes. If not, assign the random number to the current gene to ensure that the gene is not duplicate with the previous genes. The population size is controlled by the parameter scale, and scale chromosomes are generated continuously.

Finally, greedy algorithm strategy is used to realize the optimal scheduling of vehicles. Start with the customer represented by the gene generated in the random method and find the customer closest to the initial customer i As the next object of vehicle distribution, the customer i Mark as traversed, and then traverse the remaining customers that have not been traversed to find the customers that are far away i Recent Customers j, the customer j Mark as traversed, and then j As the starting customer, then find the distance customer j For the nearest customer, follow this rule to find the next customer until all customers have been traversed, and then get the full order of customers, that is, the order of vehicle delivery.

Therefore, the optimal scheduling of logistics distribution vehicles is realized.

3   Test Experiment Analysis

3.1   Test Environment Parameter Setting

(1) Time-varying road network constraint

The capacity of urban road vehicles is limited, especially the number of road lanes planned and constructed in the early stage is small, which is easy to cause congestion. The degree of traffic congestion will directly affect the speed of vehicles, and the degree of traffic congestion can be evaluated in more detail and reasonably with other easily measured or predicted traffic flow parameters, which can link other traffic flow parameters with the speed, providing a reasonable basis for the calculation of vehicle travel time. According to the relevant urban traffic management regulations, the road traffic congestion can be divided into the following four types, as shown in Table 1.

Table 1. Relationship between road traffic congestion and vehicle speed

degree of crowding	Speed (km/h)
open	≥ 30.0
Mild crowding	20.0–30.0
Crowding	10.0–20.0
Severe crowding	< 10.0

In addition, the driving speed of vehicles under the time-varying network is closely related to the road type, as well as the city size and time period. For this reason, this paper has carried out a detailed study on the relationship between the expressway and national and provincial highways in different time periods and the vehicle speed.

Analysis, and set the division criteria as shown in Tables 2 and 3.

Table 2. Relationship between Expressway and Speed in Different Time Periods

time interval	Corresponding period	Corresponding speed (km/h)
0:00–10:00	Low peak period	110
10:00–16:30	Peak period	80
16:30–24:00	Low peak period	110

Table 3. Relationship between Provincial Highway and Speed in Different Time Periods

time interval	Corresponding period	Corresponding speed (km/h)
0:00–7:00	Low peak period	60
7:00–9:30	fastigium	30
9:30–17:00	Peak period	50
17:00–18:00	fastigium	30
18:00–24:00	Peak period	60

In order to make the travel time of vehicles uniform and uninterrupted, the existing travel speed model is referred. The speed step distribution and travel time function are shown in Fig. 3 below.

It can be seen from Fig. 3 that the model meets the "first in, first out" criterion, the travel time can be described by speed, and the travel time is calculated by the ratio of link length and travel speed. In real life, although the traffic data system collects information all the time, the real-time traffic data we can know is generally updated according to a certain interval of time, such as 3min and 5min.Therefore, the data obtained in this paper can only be the data after 5 min.

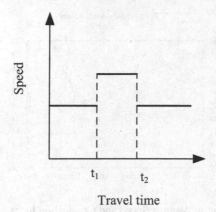

(a) Expressway running speed and running time function

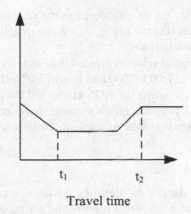

(b) Travel speed and travel time function of provincial highway

Fig. 3. Function Diagram of Travel Speed and Travel Time

(2) Actual traffic conditions.

In the face of complex urban traffic conditions, the efficiency of urban logistics distribution service will be very passive affected by it. Understanding the traffic changes in different time periods and avoiding congested roads will help the road passing rate of vehicles in transit and reduce the driving time. Therefore, this paper will collect the travel speed distribution of a certain urban road during the impasse period through relevant software. Although it is difficult to obtain specific travel directions for individual vehicles, each city basically has its own inherent general travel rules for the entire urban group. Fig. 4 below shows the trend chart of the average speed of urban roads and road congestion index in a city on a day (not a weekend holiday).

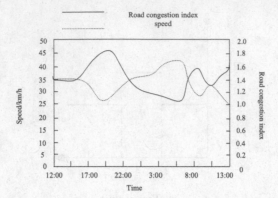

Fig. 4. Trend Chart of Average Speed and Congestion Index of Urban Roads

The road congestion index can be considered as the ratio of the actual travel time to the travel time when the road is clear. In Fig. 4, we can roughly see the peak, flat peak and valley peak periods of urban roads.

The average running speed of vehicles divided into sections is 31.2 km/h. According to Fig. 4, the time period from 7:00 to 9:00 and from 17:00 to 19:30 can be divided into peak periods, with the average speed set as 25 km/h, and the time period from 9:00 to 17:00 belongs to the flat peak period, with the average speed set as 40 km/h. The divided time period and speed are applied to the example in the next section.

3.2 Test Plan

This section will quote the data of the logistics distribution center of A logistics distribution center in a city under the actual road traffic for the experiment. The specific experimental data are shown in Table 4 below.

On this basis, considering that in the actual logistics distribution process, users' requirements for delivery time also have the characteristics of differentiation. For this reason, this paper sets time windows for different users, as shown in Table 5.

Constrain the execution of distribution tasks according to the time shown in Table 4. In the specific distribution process, the number of vehicles that can be scheduled in the corresponding area of the task is 6. Without considering the influence of other factors, the average speed of the vehicle is set to 32 km/h, and the use cost of the vehicle is 60 per vehicle. According to market statistics, the maximum single capacity of ordinary urban freight vehicles is 200 units, the energy consumption for driving is 12 units per hundred kilometers, and the energy cost per kilometer converted from the current oil price is 0.65. The logistics distribution vehicle optimal scheduling method based on genetic algorithm (control group 1) and the logistics distribution vehicle optimal scheduling method based on ant colony algorithm (control group 2) were used as the test control group to compare the distribution effect under different scheduling mechanisms.

Table 4. Test Case Detailed Data Information

Customer number	longitude	latitude	requirement
0	114. 047928	22.53404	0
1	114. 084007	22.533 106	11
2	114. 063476	22. 517449	13
3	114. 048129	22.569696	16
4	114. 117707	22.547856	9
5	114. 078438	22.557769	5
6	114. 126343	22.580903	28
7	114. 002705	22.548985	16
8	114. 150582	22.560752	14
9	114. 123136	22.566035	12
10	114.13405	22.55548	19
11	114. 10387	22. 557838	23
12	114. 089129	22.56301 1	20
13	114.005573	22.535584	8
14	114.093954	22.547285	19
15	114.05208	22. 547539	8
16	114. 033642	22.564374	12
17	114. 073548	22. 567397	8
18	113. 999491	22.542243	12
19	114. 084514	22.541893	16
20	114. 076734	22. 55464	9
21	114. 025829	22.551368	11
22	114. 051737	22. 522581	18
23	114. 070394	22.547521	29
24	114. 110004	22.570306	21
25	114. 115356	22.560368	6

3.3 Test Results and Analysis

(1) Result analysis without considering traffic conditions

Firstly, the distribution results corresponding to different test methods without considering traffic conditions are counted. The distribution duration and total distance of six logistics vehicles are shown in Fig. 5 and Fig. 6 respectively.

Table 5. Delivery Task Time Window Information

Customer number	Delivery Task Time Window
0	8:00–18:00
1	8:30–8:50
2	8:00–8:30
3	9:00–9:20
4	9:35–9:50
5	10:00–10:20
6	10:30–10:55
7	11:15–11:35
8	11:30–11:55
9	12:00–12:20
10	12:15–12:45
11	13:30–14:00
12	8:00–9:20
13	8:30–10:00
14	14:30–14:50
15	14:45–15:15
16	15:20–15:35
17	16:20–16:55
18	12:20–12:50
19	13:10–13:40
20	12:20–12:45
21	10:20–10:50
22	15:00–15:20
23	16:00–16:30
24	11:20–1150
25	10:35–10:55

According to the test results shown in Fig. 5, in the three groups of test results, the distribution duration of logistics vehicles in control group 1 and control group 2 is always higher than that in the test group. From the overall perspective, the distribution duration of six logistics vehicles in control group 1 is 1696.8 min in total, and the distribution duration of six logistics vehicles in control group 2 is 1760.4 min in total, The total delivery time of six logistics vehicles in the test group is only 1623.6 min. Compared with control group 1, the delivery time saved reached 73.2 min, and compared with control group 2, the delivery time saved reached 163.8 min. It can be seen from this that under the optimization scheduling algorithm of logistics distribution vehicles based

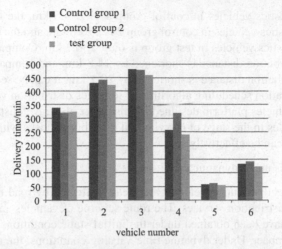

Fig. 5. Comparison Diagram of Vehicle Delivery Duration

on the Internet of Vehicles platform designed in this paper, the time cost of logistics distribution vehicles in the stage of implementing distribution tasks without considering traffic conditions can be effectively reduced. The algorithm in this article constructs an objective function that comprehensively arranges the number of vehicles, the total distance traveled by delivery vehicles, and constrains punishment with time windows. Use greedy algorithm strategy to achieve optimal vehicle scheduling. Therefore, the efficiency of vehicle delivery is good.

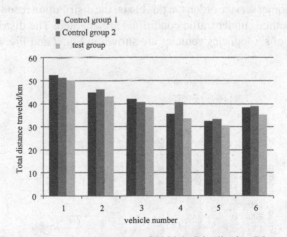

Fig. 6. Comparison Diagram of Vehicle Distribution Distance

Combined with the test results shown in Fig. 6, it can be seen that in the three groups of test results, the driving distance of logistics vehicles in control group 1 and control group 2 is always higher than that of the test group during the delivery task stage. It is also analyzed from the overall perspective, among which, the total distribution driving distance of 6 logistics vehicles in control group 1 is 245.5 km, the total distribution duration of 6 logistics vehicles in control group 2 is 250.8 km, and the total distribution duration of 6 logistics vehicles in test group is only 230.7 km. Compared with control group 1, the distribution distance is shortened by 14.8 km, and compared with control group 2, the distribution distance is shortened by 20.1 km. It can be seen from this that under the optimization scheduling algorithm of logistics distribution vehicles based on the Internet of Vehicles platform designed in this paper, the travel distance of logistics distribution vehicles in the stage of carrying out distribution tasks without considering traffic conditions can be effectively shortened.

(2) Result analysis under dynamic conditions

Urban logistics distribution vehicles are greatly affected by road traffic conditions when providing distribution services. The route scheme of vehicles and the number of service vehicles have been obtained under the initial static conditions without considering the traffic impact. Under dynamic time-varying conditions, the route of vehicles can be adjusted temporarily by predicting the traffic conditions (smooth, ordinary, congested) on the given route. The premise of adjustment is to adjust the service order or choose other road sections at the remaining unserviceable customer points of the current vehicles. Since the traffic data of the government affairs platform is not open, the road information is incomplete when using the program to obtain the historical traffic data on the corresponding road section.

The method of random congestion coefficient is used for auxiliary experiments. Randomly select a certain road section on each path for random congestion number. When the congestion number $r \leq 2$, continue to select the current path to serve the customer point. When $2 \leq r \leq 3.5$, the speed will drop to 3/4 of the original, and when $r \geq 3.5$, adjust the customer service order. On this basis, the distribution results corresponding to different test methods under traffic conditions are counted. The distribution duration and total distance of six logistics vehicles are shown in Fig. 7 and Fig. 8 respectively.

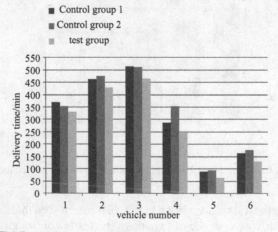

Fig. 7. Comparison Diagram of Vehicle Delivery Duration

Combined with the test results shown in Fig. 7, it can be seen that in the three groups of test results, the logistics vehicle distribution duration of control group 1 and control group 2 is always higher than that of the test group, and the range is significantly expanded compared with the test results without considering traffic conditions. From the overall point of view, the total distribution time of six logistics vehicles in control group 1 was 1884.0min, the total distribution time of six logistics vehicles in control group 2 was 4956.6min, and the total distribution time of six logistics vehicles in test group was only 163.2min. Compared with control group 1, the delivery time saved reached 220.8 min, and compared with control group 2, the delivery time saved reached 302.4 min. Moreover, compared with the test results without considering the traffic conditions, the delivery time of control group 1 increased by 187.2 min, the delivery time of control group 2 increased by 205.2 min, and the delivery time of test group only increased by 39.6 min. It can be seen from this that, considering the traffic conditions, the optimization scheduling algorithm of logistics distribution vehicles based on the Internet of Vehicles platform designed in this paper can also effectively reduce the time cost of logistics distribution vehicles in the stage of implementing distribution tasks.

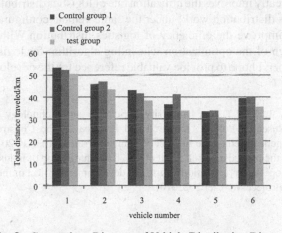

Fig. 8. Comparison Diagram of Vehicle Distribution Distance

According to the test results shown in Fig. 8, in the three groups of test results, the vehicle distribution distance shows the same development trend as the vehicle distribution duration. Among them, the total distance traveled by six logistics vehicles in control group 1 during the delivery task phase is 251.9 km, the total distance traveled by six logistics vehicles in control group 2 during the delivery task phase is 255.3 km, while the total distance traveled by six logistics vehicles in test group during the delivery task phase is only 232.1 km. Compared with control group 1, the distribution distance is shortened by 19.8 km, and compared with control group 2, the distribution distance is shortened by 23.2 km. Moreover, compared with the test results without considering traffic conditions, the distribution distance of control group 1 increased by 6.4 km, the distribution distance of control group 2 increased by 4.5 km, and the distribution distance of test group only increased by 1.4 km. It can be seen from this that, considering the traffic conditions, the optimization scheduling algorithm of logistics distribution vehicles based on the Internet of Vehicles platform designed in this paper can also effectively shorten the travel distance of logistics distribution vehicles in the delivery task phase.

4   Conclusion

Vehicle routing problem has more important practical significance because of the rapid development of logistics transportation in recent years.Many models of vehicle routing problem have been derived from the research process of scholars for many years, among which the algorithm for solving the model is endless.The models of vehicle routing problem include vehicle routing problem with load constraints, vehicle routing problem with time window constraints, dynamic vehicle routing problem, static vehicle routing problem, dynamic vehicle routing problem with time window, random vehicle routing problem, vehicle routing problem with random demands, vehicle routing problem with random customers and demands, etc.This paper proposes the research on the optimization scheduling algorithm of logistics distribution vehicles based on the Internet of Vehicles platform, which greatly improves the utilization rate of logistics distribution vehicles. For the actual logistics distribution work, under the same vehicle configuration conditions, it can effectively improve the efficiency of logistics distribution.With the help of the research and design of the optimization scheduling algorithm for logistics distribution vehicles in this paper, I hope to provide valuable reference for the development of related work.

Acknowledgement. 1. 2022 Dalian University of Science and Technology Higher Education Teaching Reform Research Project "Research and Practice on the Construction of Student Classroom Environment Based on the Internet of Things" (Project No. XJG202225).

2. This article is the phased research result of the 2022 "Integration of industry and education, school-enterprise cooperation" education reform and development project of the China Electronic Labor Association (Project No. Ciel 2022010).

References

1. Zhao, Y., Chen, Z.H.: Shortest path optimization of reverse logistics vehicle based on local search. Comput. Simul. **39**(11), 6 (2022)
2. Xu, X., Liu, L.: Research on distributed logistics scheduling method for workshop production based on hybrid particle swarm optimisation. Int. J. Manuf. Technol. Manage. **35**(3), 234 (2021)
3. Pennetti, C.A., Jun, J., Jones, G.S., et al.: Temporal disaggregation of performance measures to manage uncertainty in transportation logistics and scheduling. ASCE-ASME J. Risk Uncertainty Eng. Syst. Part A Civ. Eng. **7**(1), 04020047 (2021)
4. Ding, Z., Xu, X., Jiang, S., et al.: Emergency logistics scheduling with multiple supply-demand points based on grey interval. J. Saf. Sci. Resilience **3**(2), 10 (2022)
5. Zhang, Y., Liu, J.: Emergency logistics scheduling under uncertain transportation time using online optimization methods. IEEE Access **PP**(99), 1 (2021)

6. Ding, Z., Zhao, Z., Liu, D., et al.: Multi-objective scheduling of relief logistics based on swarm intelligence algorithms and spatio-temporal traffic flow. J. Saf. Sci. Resilience **2**(4), 8 (2021)

7. Lei, J., Hui, J., Ding, K., et al.: A framework for planning and scheduling shop floor logistics via cloud-edge collaboration. J. Phys. Conf. Ser. **1983**(1), 012109 (2021)

8. Midaoui, M.E., Qbadou, M., Mansouri, K.: Logistics chain optimization and scheduling of hospital pharmacy drugs using genetic algorithms: morocco case. Int. J. Web Based Learn. Teach. Technol. **2021**(16), 54 (2021)

9. Chen, T.: Decision-making support for transportation and logistics combining rough set fuzzy logic algorithm. J. Intell. Fuzzy Syst. Appl. Eng. Technol. **2021**(4), 41 (2021)

10. Wang, S.: Artificial intelligence applications in the new model of logistics development based on wireless communication technology. Scientific Programming **2021**(Pt.9), 2021 (2021)

11. Zhang, Q., Wang, T., Huang, K., et al.: Efficient dispatching system of railway vehicles based on internet of things technology. Pattern Recogn. Lett. **143**(6), 14–18 (2021)

12. Madahi, S.S.K., Nafisi, H., Abyaneh, H.A., et al.: Co-optimization of energy losses and transformer operating costs based on smart charging algorithm for plug-in electric vehicle parking lots. IEEE Trans. Transp. Electrification **7**(2), 527–541 (2021)

13. Li, C., Zhang, L., Zhang, L.: A route and speed optimization model to find conflict-free routes for automated guided vehicles in large warehouses based on quick response code technology. Adv. Eng. Inform. **52**, 101604–101691 (2022)

14. Masadeh, R.M.T., Alsharman, N., Sharieh, A.A., et al.: Task scheduling on cloud computing based on sea lion optimization algorithm. Int. J. Web Inf. Syst. **17**(2), 99–116 (2021)

15. Basu, M., Basu, S.: Horse herd optimization algorithm for fuel constrained day-ahead scheduling of isolated nanogrid. Appl. Artif. Intell. **35**(15), 1250–1270 (2021)

Design of Logistics Information Tracking System for Petrochemical Enterprises Under the Background of Intelligent Logistics

Bing Yuan and Yarong Zhou(⊠)

Faculty of Management, Chongqing College of Architecture and Technology, Chongqing 401331, China
15309910368@189.cn

Abstract. With the gradual expansion of the demand for petroleum and petrochemical products in China, higher requirements have been put forward for the logistics capabilities of corresponding enterprises. The items transported by petrochemical enterprises are corrosive chemicals, therefore there are higher requirements for transportation safety. In order to improve the safety of logistics transportation in petrochemical enterprises, a logistics information tracking system for petrochemical enterprises is designed in the context of smart logistics. Design the system framework structure, mainly including the foundation layer, service layer, system layer, and application layer. In the hardware part of the system, the information collected by temperature and humidity sensors is transmitted through wireless networks. In the system software section, GPS is used to locate the logistics transport vehicles of petrochemical enterprises. Design a logistics information tracking system by combining hardware and software. The experimental results show that the access delay of the designed system is relatively short, with a maximum of 1 s.

Keywords: Smart Logistics · Petrochemical Enterprises · Logistics Information · Tracking System

1 Introduction

Modern logistics, as a derivative of economic development in the Internet era, is rapidly thriving and achieving new breakthroughs and growth through continuous self revolution. Especially in the new era, science and technology are becoming more and more advanced, bringing more possibilities for the intelligent upgrading of China's logistics industry [1]. In the era of information technology, the Internet of Things (IOT) technology is a hot new generation of information technology. Through various physical recognizers, sensors, and positioning systems, combined with monitoring and object interaction, it collects information such as light, heat, and mechanics, and inputs this information into network systems to ensure certain connections between objects and people, and constantly monitors the process of object transfer and movement, At the

L. Yun et al. (Eds.): ADHIP 2023, LNICST 550, pp. 84–98, 2024.
https://doi.org/10.1007/978-3-031-50552-2_6

same time, it is also identified and managed to a certain extent [2]. In fact, the Internet of Things is a carrier based on many other systems, and it does not exist alone, let alone impossible. The foundation for the existence of the Internet of Things is still the Internet, which is an extension of the Internet. It helps users monitor any item in real-time and exchange information with the parties involved, forming close connections between people and things, as well as between things and things.

The foundation and prerequisite for the existence of smart logistics is the Internet of Things. Smart logistics is the use of the latest information processing technology to centralize information related to logistics, using a central processor to process data, and selecting the optimal solution from all options to assist logistics operations [3]. This approach is both efficient and flexible, meeting people's logistics needs and being able to respond to the ever-changing logistics situation at the fastest speed. Intelligent logistics utilizes intelligent methods to influence the logistics transportation process. Perception technology is a relatively complex technology that requires the perception of multiple basic requirements such as the carrier, flow base, and flow mode of objects in the logistics process, making the logistics operation process more convenient and fast. Real time monitoring of the logistics transportation process, establishing transmission platforms and technical platforms, and realizing the mutual transmission of information between logistics transportation entities, making information exchange more convenient.

With the increasing demand for logistics development from social development, China has launched policies, guidelines, and plans to support the development of smart logistics. The country takes various excellent projects launched by IoT enterprises as demonstration projects, in order to drive other enterprises to deeply explore the field of smart logistics. With the increasing application range of Internet of Things technology, various enterprises and people from all walks of life have an increasing understanding of smart logistics, which has accelerated the development speed of smart logistics and strengthened its foundation [4]. The Internet of Things technology is not only applicable to the logistics field, but can also be applied to medical devices, medical drugs, agricultural and sideline products, food and other fields, helping people track the origin and shipping process of items when shopping. China's smart logistics technology is gradually maturing, with strong policy support and technical support. Many enterprises have established their own intelligent logistics distribution centers to make the logistics operation more intelligent and automated, so that the logistics industry and the production field interact with each other, and realize the overall promotion of the flow of goods, goods and information..

Reference [5] proposes a logistics information tracking system based on the Internet of Things technology. In the hardware part, wireless sensors are designed and Kalman filters are processed. In the software part, GPS logistics information is collected, logistics information is screened, tracking authority of logistics database is identified, Internet of Things tracking model is established and information tracking is carried out. Reference [6] proposes a logistics information tracking system based on blockchain, which takes the blockchain cloud platform as the core of the system application and connects the logistics information management module and information service module respectively in the system. On this basis, the obvious characteristics of logistics users are analyzed, and all data parameters are transferred to the cloud database to facilitate the query of system application data. The security of relevant logistics information in the system is ensured by the asymmetric information encryption.

Aiming at the particularity of logistics transportation in petrochemical enterprises, a new intelligent logistics information tracking system for petrochemical enterprises is proposed and designed under the background of intelligent logistics.

2   Architecture of Logistics Information Tracking System for Petrochemical Enterprises

The logistics information tracking system focuses on the main supply chain of "production, warehousing, transportation, distribution, and sales", tracking the entire process of goods from production to sales, and even consumption, in order to provide services to different users. Individual users can track and query their purchased products through the logistics information tracking system application, and obtain information for anti-counterfeiting inspection of goods; Suppliers and sellers can adjust their production and sales plans in a timely manner through the logistics information tracking system; Logistics enterprises adjust their warehousing and distribution plans in a timely manner through tracking systems; Government agencies can supervise goods through logistics tracking systems [7–9]. According to the requirements of multiple users and links, a logistics information tracking system model is constructed as shown in Fig. 1. This system adopts a four layer architecture of basic layer, service layer, system layer, and application layer.

The basic layer consists of two parts: a hardware support platform and an EPC network system. The hardware support platform consists of a server system, storage system, and disaster recovery system. The server system mainly provides server operation support for the operation of logistics information tracking systems. The storage system mainly stores various types of data. The disaster recovery system can ensure the integrity of the logistics information tracking system in case of sudden problems. The EPC network system mainly includes three parts: network system, data collection system, and tracking goods [10–13]. The network system is mainly composed of INTERNET, wireless network, GPS, etc. The data collection system consists of electronic tags, readers, PDAs, barcode readers, and other data collection equipment. By establishing the basic layer system, provide the underlying data source for the logistics information tracking system.

The service layer includes directory services, retrieval and tracking services, information distribution services, workflow services, etc. Directory services are used to discover, query, and access RFID related data information; The retrieval service can retrieve all relevant product information based on the EPC code, and organize feedback according to user needs; Tracking service collects all process information of goods throughout their lifecycle, organizes and provides tracking, traceability, anti-counterfeiting and other services; The information distribution service adopts a publish/subscribe method to provide users with product related composite event information; The workflow service system realizes the interaction between service layer, basic layer data source, and application layer users, achieving business automation [14].

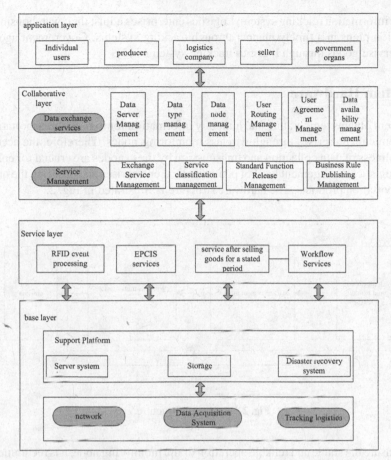

Fig. 1. Architecture of Logistics Information Tracking System

The collaboration layer includes data exchange service management and service management. The establishment of data exchange service management is to unify the basic parameters of each data exchange server for application layer users, manage each data server, data type, data node, user routing, user protocol, data availability, etc., and achieve synchronization of data changes in each data exchange service center. Service management includes exchange service management, service classification management, standard function publishing and management, and business rule publishing and management. Service management provides information service intermediaries for the service layer and application layer.

, The application layer is an application system developed for users. Through the application layer, individual users, manufacturers, logistics enterprises, sellers, and government agencies can track and query logistics information. Individual users can track and query their purchased products through the logistics information tracking system application, and obtain information for anti-counterfeiting inspection of goods; Suppliers and sellers can adjust their production and sales plans in a timely manner through the logistics information tracking system; Logistics enterprises adjust their warehousing and distribution plans in a timely manner through tracking systems; Government agencies can supervise goods through logistics tracking systems [15].

3   System Hardware

The system obtains temperature and humidity information for logistics transportation in petrochemical enterprises through wireless monitoring nodes. Therefore, the accuracy and timeliness of data collection and transmission by these nodes are crucial for enhancing the logistics management level of petrochemical enterprises and ensuring the quality of transportation products. The hardware structure is illustrated in Fig. 2.

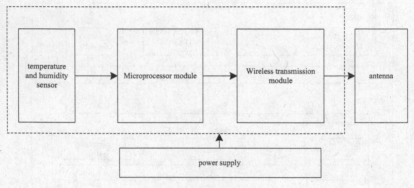

Fig. 2. Hardware Structure

To ensure the stable and reliable operation of the monitoring node, it is recommended to supply the necessary power through a mains transformer since the node is placed inside the transport vehicle.

The temperature and humidity sensor adopts the digital temperature and humidity sensor chip SHT11 from Sensirion, Switzerland. Its characteristic is that the functions of temperature sensing, humidity sensing, signal conversion, A/D conversion, and heater are high-precision integrated into one chip, providing a two wire digital serial interface SCK and DATA, with a simple interface, supporting CRC transmission verification, and high transmission reliability; Programmable adjustment of measurement accuracy with built-in A/D device; High measurement accuracy, providing temperature compensated humidity measurement values and high-quality calculation functions; The packaging size is extremely small, and after measurement and communication, it automatically switches to low-power mode; High reliability, using CMOS technology, the sensing head can be completely immersed in water during measurement; The temperature measurement range is −40 °C ~ 128.3 °C, and the humidity measurement range is 0–100% RH; SHT11 adopts SMD (LCC) surface mount packaging, with a very simple interface. The pin names and functions are: pin 1 and pin 4- signal ground and power supply, with a working voltage range of 2.4–5.5 V; Pin 2 and Pin 3- Two wire serial digital interface, where DA-TA is the data line and SCK is the clock line; Pin 5- Pin 8- Not connected.

The microprocessor module adopts LM3S811, supporting an ARM Cortex-M3 core with a maximum main frequency of 50 MHz, 64 Kbyte FLASH, 8 Kbyte SRAM, and LQFP-48 packaging. Integrated orthogonal encoder, 4-way 10 bit ADC, PWM with dead band, analog comparator, 3 universal timers, 2-way UART, SPI, I2C bus interfaces, etc. In addition to normal working mode, it also has sleep and deep sleep modes, which is very suitable for low energy applications and fully meets the requirements of Zigbee protocol stack for microprocessors.

The wireless transmission module adopts Chipcon's first CC2420 RF transceiver that meets the 2.4 GHz IEEE802.15.4 standard, which is the first RF device suitable for Zigbee products. It is based on Chipcon's SmartRF 03 technology and is made with a 0.18 um CMOS process, requiring minimal external components, stable performance, and extremely low power consumption. Adopting O-QPSK debugging method, high reception sensitivity, 4-bus SPI interface, QPL-48 packaging, with an overall size of only 7 mm × 7 mm, MAC layer hardware can support automatic frame format generation, synchronous insertion and detection, 16bit CRC verification, power detection, and fully automatic MAC layer security protection. The selectivity and sensitivity index of CC2420 exceed the requirements of the IEEE802.15.4 standard, ensuring the effectiveness and reliability of short distance communication. The wireless communication equipment developed using this chip supports a data transmission rate of up to 250 kbps.

The microcontroller LM3S811 communicates with SHT11 through a two wire serial digital interface, and the hardware interface circuit is simple, as shown in Fig. 3.

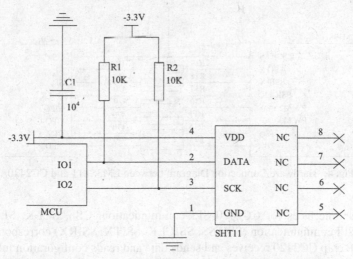

Fig. 3. Hardware Connection Diagram between LM3S811 and SHT11.

The communication protocol is not compatible with the universal I2C bus protocol, so it is necessary to simulate the communication program using a universal microprocessor I/O port. The control of SHT11 by the microprocessor is achieved through five 5-bit command codes, and the meaning of the command codes is shown in Table 1.

Table 1. Command Codecommand

command	code
temperature measurement	00011
Humidity measurement	00101
Read Status Register	00111
Write Status Register	00110
Software Reset	11110

LM3S811After sending the read status register command 0x07 to SHT11, the DATA port of SHT11 transmits temperature and humidity data to LM3S811.

The communication interface connection circuit between the CC2420 RF chip in the wireless transmission module and the microcontroller LM3S811 is shown in Fig. 4.

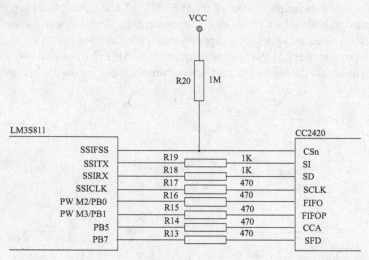

Fig. 4. Hardware Connection Diagram between LM3S811 and CC2420

The four interfaces for CC2420 SPI communication: CSn, SCLK, SI, SO, and LM3S811 SPI communication (SSIFSS, SSICLK, SSITX, SSIRX) correspond one-to-one. The RF chip CC2420 receives and sends data and reads configuration information through these lines controlled by LM3S811. CC2420 operates in slave mode, while LM3S811 operates in master mode.

4   Systems Software

GPS satellite constellation, ground monitoring part and GPS receiver constitute the GPS system, as shown in Fig. 5 below.

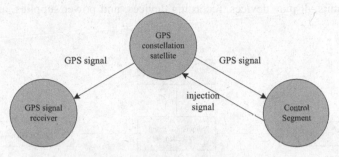

Fig. 5. GPS System Composition

(1) Space Constellation Part.

The space constellation of the Global Positioning System includes a total of 24 satellites, including 3 backup satellites. These 24 satellites are evenly distributed on six orbital planes with an average altitude of approximately 20200 km, and their operation time for one week is approximately 11:58. Each orbit of all satellites is approximately circular, and the maximum eccentricity is about 0.01. In each orbit plane, there are four satellites with an interval of 90°, and their operation period is 12 h. GPS receivers at the same location on Earth display the same satellite distribution pattern every day, with the difference being approximately 4 min earlier in time. So, if conditions permit, each point on Earth can receive signals from at least 4 satellites, and at most 11 satellites, which can meet the needs of positioning and navigation.

(2) Ground monitoring part.

The ground monitoring part of the global positioning system consists of a main control station, three information injection stations, and five monitoring and tracking stations. The main control station, also known as the Joint Space Execution Center (CSOC), is responsible for collecting data from various monitoring stations, generating navigation messages, transmitting them to the injection station, and providing a time reference. In addition, the main control station can also control and manage the work of each monitoring station and injection station. When the satellite deviates, adjustments can be made to ensure the normal operation of the entire system. The job responsibility of the injection station is to inject the prepared navigation messages into the storage space of each satellite, improving the accuracy of the satellite's broadcast signal. Each monitoring station is equipped with dual frequency GPS receivers and high-precision cesium clocks, which can track and measure satellites in real-time and transmit the collected information to the main control station at any time.

(3) Ground receiving equipment.

The GPS satellite constellation and ground monitoring described above are indispensable for the user to carry out positioning and navigation, but positioning and navigation cannot be realized without a receiver. The selection of GPS receivers varies due to different needs and environments, but general GPS receivers include antennas, signal processing units, display devices, recording devices, and power supplies, as shown in Fig. 6.

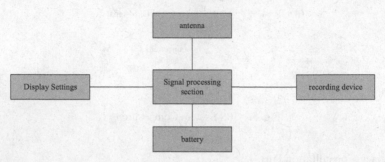

Fig. 6. Basic Structure of GPS Receiver

The principle of GPS positioning is mainly based on geometric and physical principles, utilizing the distribution of space satellites and the distance between satellites and ground points to intersect and determine the specific location of ground points. Assuming that we already know the position of the satellite and can accurately determine the distance between observation point A and the satellite through special measurement methods, we can accurately infer that point A must be located on a sphere centered around this satellite and with the measured distance as the radius. Similarly, if we can also know the distance between point A and any two other satellites, then the position of observation point A must be at the intersection of three spheres.

However, the positioning process for GPS satellites involves measuring three distances simultaneously, as the satellites are distributed beyond 20,200 km. To achieve synchronization and accurate positioning, it is crucial to have a common reference time. According to analytical geometry, GPS positioning necessitates the knowledge of the three-dimensional coordinates of a point and the synchronization of four unknown quantities. Therefore, a minimum of four satellite distances must be measured for precise positioning, as depicted in Fig. 7.

The actual distance R from the satellite to the receiver is calculated by measuring the time Δt from the GPS signal to the receiver. The specific calculation formula is:

$$R = c\Delta t \tag{1}$$

In the formula, c represents the propagation speed of the GPS signal.

Due to the influence of clock difference, the receiver's time V_{t0} and UTC's time V_{ti} may not necessarily be synchronized, i.e. $V_{ti} - V_{t0} \neq 0$. When UTC is fast, then $V_{ti} - V_{t0}$ is positive, otherwise it is negative. Due to the existence of time error $V_{ti} - V_{t0}$, there

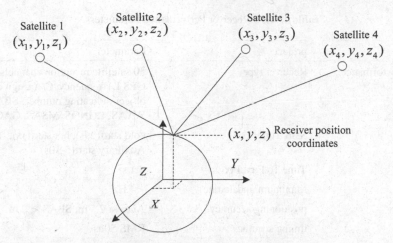

Fig. 7. Receiver position coordinate calculation

is an error in the signal transmission distance. Therefore, the measured distance to the satellite is called pseudo distance ρ, and the calculation formula is as follows:

$$\rho = R + c(V_{ti} - V_{t0}) \tag{2}$$

Calculate the coordinates of the four satellites based on the ephemeris file, and let $(x_i, y_i, z_i)(i = 1, 2, 3, 4)$ represent the spatial indicator coordinate systems of the four satellites at time t. Based on the product of the time Δt of signal transmission and the speed of light, the distance between the satellite and the receiver, known as pseudo distance, can be obtained. Let $\rho_i(i = 1, 2, 3, 4)$ represent the distance between the four satellites and the receiver, and $\Delta t_i(i = 1, 2, 3, 4)$ represent the time it takes for the signals from the four satellites to reach the receiver. At the same time, suppose that $V_{ti}(i = 1, 2, 3, 4)$ represents the satellite clock difference of four satellites respectively, which is provided by the satellite ephemeris, and V_{t0} is the receiver clock difference. The receiver position (x, y, z) can be solved by the following Equation solving, so as to determine the location of the petrochemical enterprise logistics transport vehicle. The specific calculation formula is:

$$
\begin{aligned}
{}^{1/2} + c(V_{t1} - V_{t0}) &= \rho_1 \\
[(x_2 - x)^2 + (y_2 - y)^2 + (z_2 - z)^2]^{1/2} + c(V_{t2} - V_{t0}) &= \rho_2 \\
[(x_3 - x)^2 + (y_3 - y)^2 + (z_3 - z)^2]^{1/2} + c(V_{t3} - V_{t0}) &= \rho_3 \\
[(x_4 - x)^2 + (y_4 - y)^2 + (z_4 - z)^2]^{1/2} + c(V_{t4} - V_{t0}) &= \rho_4
\end{aligned}
\tag{3}
$$

5   System Performance Verification

Propose a comparative experiment, using the reference [5] system as a comparative system to compare the tracking performance of this system and the comparative system when facing the same logistics information of petrochemical enterprises.

The performance parameters of the GPS receiver are shown in Table 2.

Table 2. GPS Receiver Performance Parameters

	project	Parameter
Basic performance	Receiver type	50 satellite reception channels, GPS L1frequency, C/Aa sign or object indicating number; SBAS: WAAS, EGNOS, MSAS, GAGAN
	Start Time	cold start(29s), hot start(1s), Auxiliary start(5-10s)
	Time To First Fix	<1 s
	Maximum update rate	<4 Hz
	positioning accuracy	Auto < 2.5 m, SBAS < 2 m
	timing accuracy	RMS 50ns
	Antenna power supply method	External or InternalVCC-RF
	Antenna status detection	Integrated short circuit detection and antenna shutdown function; Open circuit detection is provided by AADET-N input and external small circuit
	Ultimate operating speed	515 m/s
	Store Wendy	−40~85 °C
	Package Size	17 × 22.4 mm
Electrical performance	working voltage	2.7 V~3.6 V
	consumption	120 mW@3.0 V
	Spare battery	1.3 V~4.8 V, 30 uA
	I/Olevel	3 V
interface protocol	serial interface	1 UART, 1 I2C USB V2.0 (full speed12Mbit/s)
	Other interfaces	Full speed 1 time pulse output, 1 external interrupt input, 1 reset pin, 1 configuration pin
	agreement	NMEA, UBXbinary system

5.1 System Access Delay

The comparison results of access latency between the two systems are shown in Table 3.

From the system access delay results shown in Table 3, it can be seen that compared to traditional systems, the access delay of this system is significantly shortened, with a maximum of no more than 1 s. Therefore, it indicates that the system in this article has good operational performance and fast access speed. Low access latency is crucial for real-time data acquisition and response. In the field of logistics and transportation, timely access to temperature and humidity data is crucial, as these data play an important role in

Table 3. System Access Delay

Number of experiments	Access latency/s	
	This article system	Traditional systems
1	0.63	5.61
2	0.65	6.64
3	0.69	3.69
4	0.70	5.73
5	0.72	6.85
6	0.74	6.71
7	0.68	5.87
8	0.69	4.85
9	0.80	4.88
10	0.72	4.89

ensuring the quality and safety of goods. This system can efficiently collect, process, and transmit data, ensuring timely access to temperature and humidity information, thereby improving the level of logistics management. In addition, shorter access latency can also improve the user experience. Whether it's internal logistics management personnel, customers or partners, they can quickly obtain the necessary data and make more accurate decisions. This helps to improve the efficiency and smoothness of the entire supply chain process.

In summary, by shortening access latency and achieving faster access speed, the system proposed in this article has good operational performance in the field of logistics and transportation, and can meet the needs of real-time data acquisition and user experience.

5.2 Logistics Information Tracking Effect

Import the logistics information of a certain petrochemical enterprise in the past 10 years into two testing systems, randomly select six logistics information from them, and use two systems to track and set six logistics information in the nearly 2.65TB network data. The basic information of the testing objects is shown in Table 4.

Import all information into the system database and use the logistics information tracking programs of two systems to identify the logistics information in Table 4. In order to make the experimental results more reliable, a total of 10 experiments were conducted. The test results of a random group are shown in Tables 5 and 6.

Table 4. Logistics Information to be Tracked

type of goods	Logistics vehicle number	Order number	Delivery time	Unloading warehouse
A1-15	S1	KBU8546	2012-4-12	C03-1
A3-07	S2	KAU8744	2016-8-7	C02-4
B1-02	S3	KFU0759	2018-7-11	C03-5
F4-13	S4	KDU2434	2010-5-8	C03-3
J2-06	S5	KDU1885	2015-12-25	C02-1
H5-08	S6	KAU579720	2018-9-18	C01-3

Table 5. Tracking results of the system in this article

type of goods	Logistics vehicle number	Order number	Delivery time	Unloading warehouse
A1-15	S1	KBU8546	2012-4-12	C03-1
A3-07	S2	KAU8744	2016-8-7	C02-4
B1-02	S3	KFU0759	2018-7-11	C03-5
F4-13	S4	KDU2434	2010-5-8	C03-3
J2-06	S5	KDU1885	2015-12-25	C02-1
H5-08	S6	KAU579720	2018-9-18	C01-3

Table 6. Tracking Results of Traditional Systems

type of goods	Logistics vehicle number	Order number	Delivery time	Unloading warehouse
A3-07	S2	KAU8744	2016-8-7	C02-4
B1-02	S3	KFU0759	2018-7-11	C03-5
J2-06	S5	KDU1885	2015-12-25	C02-1
H5-08	S6	KAU579720	2018-9-18	C01-3

According to the test results shown in Tables 5 and 6, it can be seen that the system in this paper obtained all the logistics information of the six test objects, while the traditional system did not track the logistics information before 2015, indicating that the performance of the system in this paper is stronger. The system proposed in this article exhibits stronger performance and functionality in obtaining and tracking logistics information, providing enterprises with more comprehensive and accurate logistics data, thereby improving the level and effectiveness of logistics management.

For logistics management, obtaining and tracking complete logistics information is crucial for enterprise decision-making and operation. Through the application of this system, enterprises can have a comprehensive understanding of various links and data in the logistics process, including starting points, transit stations, transportation routes, arrival times, etc. This information helps enterprises to carry out precise logistics planning, optimize transportation routes, improve transportation efficiency, and better respond to potential problems and risks.

6 Conclusion

In this study, a system framework structure was designed to address the logistics information tracking needs of petrochemical enterprises in the context of smart logistics. This framework includes the basic layer, service layer, system layer, and application layer, providing a comprehensive and efficient logistics information management solution for petrochemical enterprises. In the hardware part of the system, a wireless network is introduced to transmit the information collected by temperature and humidity sensors. This enables real-time monitoring and recording of temperature and humidity data during the logistics process, thereby ensuring the quality and safety of goods during transportation. Through wireless network transmission, data can be quickly and accurately obtained and transmitted, improving the efficiency of logistics information collection and transmission. In the system software section, GPS positioning technology is used to track the location of logistics transportation vehicles in petrochemical enterprises. This enables enterprises to real-time grasp the driving situation and location of vehicles, which helps optimize route planning, improve transportation efficiency, and respond to potential problems and risks in a timely manner.

By combining hardware and software, a fully functional, efficient and reliable logistics information tracking system has been successfully designed. This system can help petrochemical enterprises achieve visualization and intelligent management of logistics processes, improve the efficiency and quality of logistics transportation, and reduce costs and risks. In summary, this study provides an effective solution for the design of logistics information tracking systems for petrochemical enterprises in the context of smart logistics. I believe that the application of this system will bring higher logistics management level and competitive advantage to petrochemical enterprises, and contribute to the sustainable development of the industry. In the future, the system will be further optimized and improved to meet constantly changing needs and challenges.

Aknowledgement. Research Achievements of the Digital Innovation Research Team of Chongqing College Of Architecture And Technology(Project No.: CXTD22B01).

References

1. Yang, F.: E-Commerce logistics system based on internet of things. J. Interconnection Netw. **22**(3), 214–219 (2021)
2. Zu, E., Shu, M.H., Huang, J.C., et al.: Management problems of modern logistics information system based on data mining. Hindawi Limited (2021)

3. Sun, J., Fan, Y.: Research on E-commerce logistics information system based on big data technology. J. Phys. Conf. Ser. **1972**(1), 120–128 (2021)
4. Deng, W., Maki, A.: Design of logistics transportation monitoring system based on GPS/DR combined positioning technology. Int. J. Inf. Technol. Manage. **20**(3), 282–298 (2021)
5. Yan, Y.: Real time tracking system for storage and transportation logistics information based on Internet of Things technology. Chin. Storage Transp. **09**, 72–73 (2021)
6. Chen, L., Feng, Q.: Application research of third-party logistics information real time update system based on blockchain. Electron. Des. Eng. **29**(22), 42–46 (2021)
7. Gu, J.C., Yao, C., Jiang, T.H.: ELTIML: Express logistics tracking information markup language for data exchange processes in express logistics. J. Comput. Methods Sci. Eng.Comput. Methods Sci. Eng. **21**(2), 397–407 (2021)
8. Midaoui, M.E., Laoula, E., Qbadou, M., et al.: Logistics tracking system based on decentralized IoT and blockchain platform. Indonesian J. Electric. Eng. Comput. Sci. **23**(1), 421–430 (2021)
9. Kuang, M., Du, Y., Kuang, D., et al.: Research on the optimization of urban cold chain logistics system based on the example of Guangzhou. J. Phys. Conf. Ser. **174**(6), 127–135 (2021)
10. Adekola, O.D., Udekwu, O.K., Saliu, O.T., et al.: Object tracking-based "follow-me" unmanned aerial vehicle (UAV) system. Comput. Mater. Continua **36**(3), 342–348 (2022)
11. Wang, H., Li, X.: Design of logistics information monitoring system based on data mining. Electron. Des. Eng. **30**(06), 71–75 (2022)
12. Li, Z., Zhao, N., Yin, M., et al.: Construction of port and shipping logistics information system based on production berth resource sharing. J. Transp. Syst. Eng. Inf. Technol. **23**(01), 275–283 (2023)
13. Wu, N., Dai, H., Li, J., et al.: Multi-objective optimization of cold chain logistics distribution path considering time tolerance. J. Transp. Syst. Eng. Inf. Technol. **23**(02), 275–284 (2023)
14. Zhao, J.: Design of low carbon logistics distribution center location optimization system based on improved ant colony algorithm. Mod. Electron. Tech. **44**(13), 96–100 (2021)
15. An, S., Zhao, Y.: Research on positioning method of logistics storage system based on RFID technology. Inf. Technol. **11**, 156–161 (2021)

Real Time Tracking of the Position of Intelligent Logistics Cold Chain Transportation Vehicles Based on Wireless Sensor Networks

Dong'c Zhou[1](✉) and Xunyan Bao[2]

[1] Guangzhou Huashang Vocational College, Guangzhou 511300, China
zhoudonge52413@163.com
[2] Zhejiang Changzheng Vocational and Technical College, Hangzhou 310012, China

Abstract. In order to more effectively track intelligent logistics cold chain transportation vehicles in real-time, This article proposes a real-time tracking algorithm for the position of intelligent logistics cold chain transportation vehicles based on wireless sensor networks. In the first stage, the received RSSI value of the anchor node is directly used for coarse positioning. In the second stage, the coarse positioning value is used as the initial solution optimization iteration of the wireless sensor network. In addition, this article also conducts research on the prediction and tracking of location nodes for smart logistics cold chain transportation vehicles, and discusses real-time tracking algorithms for EKF and wireless sensor networks. In view of the fact that EKF algorithm needs to abandon the information of higher order items of the system when approaching the nonlinear, which has caused error accumulation to some extent, the real-time tracking algorithm of wireless sensor network uses UT transformation and deterministic sampling strategy to linearly map the nonlinear system, which retains the information of the system to the greatest extent, and realizes effective prediction and real-time tracking of the nonlinear system.

Keywords: Wireless Sensor Network · Smart Logisticsm · Cold Chain Transportation · Vehicle Position Tracking

1 Introduction

In this paper, the wireless sensor network is used to track the location of intelligent logistics cold chain transport vehicles in real time, so as to timely understand the location information of cold chain transport vehicles. The smart logistics cold chain transport vehicle location real-time tracking method based on wireless sensor network has the characteristics of miniaturization and low price [1]. They have the characteristics of wireless communication and AD hoc network, which makes the smart logistics cold chain vehicle location real-time tracking method based on wireless sensor network has significant advantages over the traditional tracking method, which is reflected in the following aspects: (1) more precise tracking. Due to the dense deployment of wireless

L. Yun et al. (Eds.): ADHIP 2023, LNICST 550, pp. 99–112, 2024.
https://doi.org/10.1007/978-3-031-50552-2_7

sensor network nodes and the characteristics of deeply embedded environment, it can realize the accurate perception, tracking and control of moving targets in close range, so as to obtain more detailed information about targets. (2) tracking can be opened. Wireless sensor networks have the characteristics of self-organization, self-configuration and intensive deployment, which makes the target tracking method of wireless sensor networks have higher reliability, fault tolerance and robustness. (3) Distributed tracking. The distributed nature of wireless sensor network makes real-time tracking based on its position without calculation and control, and reduces the risk of single point failure. It has the characteristics of small tracking delay and good scalability.

Wireless sensor network plays a huge advantage in the real-time tracking method of cold chain transport vehicle position in intelligent logistics, and has a better prospect and future because of its little overhead, low error and targeted data measurement ability.

2   Intelligent Logistics Cold Chain Transport Vehicle Location Real-Time Tracking Algorithm

2.1   Cost Analysis of Intelligent Logistics Cold Chain Transportation

There are many inevitable factors and unexpected situations in smart logistics cold chain transportation, such as weather conditions, traffic conditions, and sudden vehicle damage [2, 3]. These influencing factors have a certain impact on the transportation process in reality.

(1) Vehicle fixed cost.

Fixed costs include vehicle maintenance and repair costs, driver salaries, etc., which are necessary costs for using the vehicle. The formula for fixed cost is:

$$C_1 = \sum_{k=1}^{K} f_k \tag{1}$$

where, C_1 represents the Fixed cost of transport vehicles; f_k represents the fixed cost of car k; K represents the number of cold chain delivery vehicles, K represents the vehicle serial number $(1 \cdots k)$.

Driving cost refers to the cost of fuel consumption and passing expenses incurred during the transportation of goods. The formula of driving cost is as follows:

$$C_2 = \sum_{k=1}^{K} \sum_{i=1}^{N} \sum_{j=1}^{N} \alpha d_{ij}^k X_{ij}^k \tag{2}$$

Among them,

$$X_{ij}^k = \begin{cases} 1(\text{Cold chain delivery vehicle k from customer i to j}) \\ 0(\text{Cold chain delivery vehicle k does not go from customer i to j}) \end{cases}.$$

Where, C_2 represents the total transportation cost of all cold chain transportation vehicles; α represents the transportation cost per unit of transportation distance generated by transportation vehicles; d_{ij}^k represents the distance traveled by the k vehicle during the delivery process from customer i to j; N represents the customer Total Collection $N = \{i, j / i, j = 1, 2, \ldots, N\}$; X_{ij}^k represents the evaluation coefficient.

(2) refrigeration cost.

Cold chain transport vehicles will transport goods due to the particularity of transport products, vehicles in the process of driving and loading and unloading must ensure the quality of transport goods, because the goods must be in low temperature conditions to maintain product safety and quality, so each cold chain transport vehicle will use a certain fuel consumption to ensure the temperature state, resulting in a certain refrigeration cost, refrigeration cost formula is as follows:

$$C_3 = \sum_{k=1}^{K} \sum_{j=1}^{N} X_j^k Q_j \beta t_{ij} \tag{3}$$

Among them, $X_J^K = \begin{cases} 1(\text{The k vehicle serves customer j}) \\ 0(\text{The k vehicle does not serve customer j}) \end{cases}$.

Where, C_3 represents the total cooling cost generated by all vehicles in this experiment; β represents the refrigeration oil consumption price per unit product per unit time; t_{ij} represents the time for transportation vehicles from i to j; Q_j represents the total amount of fresh agricultural products ordered by customer j ($j = 1, 2, \ldots, N$); X_J^K represents the evaluation coefficient.

(3) Cost of goods loss.

Freight damage cost is an important cost expenditure in the process of cold chain transportation. Due to the characteristics of cold chain products, there is a certain loss rate of products due to temperature changes in the process of transportation and loading and unloading [4]. The formula of freight damage cost is as follows:

$$C_4 = \sum_{k=1}^{K} \sum_{j=1}^{N} pq_j X_J^K e^{-\phi - t_{jk}} \tag{4}$$

Among them, $X_J^K = \begin{cases} 1(\text{The k vehicle serves customer j}) \\ 0(\text{The k vehicle does not serve customer j}) \end{cases}$.

Where, C_4 represents the total cost of goods damage incurred by all delivery vehicles during the delivery process in this experiment; p represents the price of Cold Chain Goods; q_j represents the customer j's demand for goods ($j = 1, 2, \ldots, N$); ϕ represents the sensitivity coefficient of the quality of cold chain goods to time; t_{jk} represents the time when the k transport vehicle arrived at customer point j.

(4) Time window penalty cost.

Penalty cost refers to the penalty cost incurred when the delivery vehicle fails to arrive at the distribution point in time according to the customer's time requirements [5].

The formula of penalty cost is:

$$C_5 = P\left(t_j^k\right) = \begin{cases} M, t_j^k < t_1 \text{ or } t_1 > t_4 \\ \theta_1\left(t_1 - t_j^k\right), t_1 < t_j^k < t_2 \\ 0, t_2 \leq t_j^k \leq t_3 \\ \theta_2\left(t_j^k - t_3\right), t_3 < t_j^k < t_4 \end{cases} \tag{5}$$

where, C_5 represents the total penalty cost of all transportation vehicles during the delivery process in this experiment; t_1 represents the earliest time that customers can accept vehicles transporting goods to reach the delivery point; t_2 represents the earliest time the customer requires the vehicle to deliver the goods to reach the delivery point; t_3 represents the customer requests the latest time for the vehicle transporting the goods to reach the delivery point; t_4 represents the latest time that customers can accept vehicles transporting goods to reach the delivery point; θ_1 represents the penalty coefficient for vehicles transporting goods reaching the delivery point is between t_1 t_2; θ_2 represents the penalty coefficient for vehicles transporting goods reaching the delivery point is between t_3 t_4; θ_1, θ_2 is the parameter, and its value depends on customers' demand for delivery time; M is a larger positive number. The time outside t_1 t_4 refers to the time when the customer does not accept the goods. The arrival of the vehicle within this interval requires paying a large penalty fee or being rejected by the customer; t_2 t_3 is the customer's ideal distribution time range. When the delivery vehicle arrives within this period, no penalty fee will be paid; t_1 t_2 and t_3 t_4 are the time range within which the customer can accept the delivery of the vehicle, subject to the payment of penalty charges. When the delivery vehicle arrives within the t_1 t_2 range (earlier than the ideal delivery time), it will not have a significant impact on product quality and sales, and the impact factor is relatively small; When the delivery vehicle arrives within the t_3 t_4 range (later than the ideal delivery time), it has a significant impact on the freshness of the goods and has a significant impact factor.

2.2   Real Time Tracking of Vehicle Position Nodes in Wireless Sensor Networks

The first step is to measure the distance. Distance measurement is the basis of real-time tracking algorithm for intelligent logistics cold chain transport vehicles based on wireless sensor network [6–8]. The accuracy of distance measurement will directly affect the accuracy of positioning results of such algorithms. In practical application, there are many methods for distance measurement [9]. At present, the commonly used methods include Time of arrival (TOA), Time Difference of arrival (TDOA), There are three Received signal strength indications (RSSI).

(1) Time of Arrival (TOA).

The basic principle of this method is that the sending node sends the ranging signal to the receiving node, and the receiving node uses the transmission time and speed of the signal to calculate the distance between itself and the sending node. TOA is currently one of the most widely used ranging methods [10, 11]. The commonly used ranging signals include acoustic signals and radio frequency signals. The TOA method has high ranging accuracy, but it requires precise clock synchronization and high-precision clock frequency from both the transmitter and receiver, thus requiring high hardware requirements for sensor nodes [12]. Recently, due to the emergence and development of UWB technology, TOA technology has broader application prospects.

(2) Arrival time difference (TDOA).

In the TDOA ranging mechanism, the sending node simultaneously sends two wireless signals with different propagation speeds, and the receiving node calculates the distance from the sending node to itself according to the time difference between the arrival of the two signals and the propagation speed of the two signals [13–15]. This works as shown in Fig. 1.

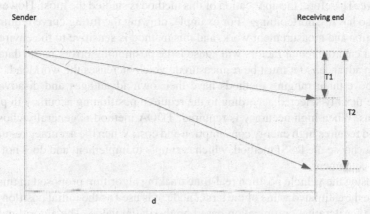

Fig. 1. TDOA measurement principle diagram

The distance between two nodes can be calculated according to (6):

$$d = (T_2 - T_1)\frac{v_1 v_2}{v_1 - v_2} \qquad (6)$$

Among them, T_1, T_2 and v_1, v_2 represent the transmission time and speed of the two signals, respectively.

The ranging accuracy of TDOA technology is higher than that of TOA method, but additional hardware equipment must be installed at the node. In practical applications, the two types of signals commonly selected for transmission are electromagnetic waves and ultrasonic waves. Nodes must install ultrasonic transceivers, which greatly increases the cost and power consumption of nodes. In addition, ultrasonic signals are easily affected by the environment, and humidity, temperature, and wind speed can all affect the accuracy of measurement.

(3) Received signal strength (RSSI).

The idea of RSSI ranging is that the receiving node measures the power intensity of the received signal and calculates the distance information according to the strength of the received signal according to the path loss model. The calculation formula is as follows:

$$d = d_0 * 10^{(RSS_0 - RSS)/10n_p} \tag{7}$$

Among them, represents the signal strength received by RSS, RSS_0 represents the signal strength after transmitting a reference distance d_0, and n_p is the path loss index, which varies according to the surrounding environmental conditions and is generally taken as 2–4.

The strength of the received signal can be obtained directly by the node without additional hardware equipment. At the same time, the method requires the least calculation in the actual ranging process, and the distance estimate can be obtained only by looking up the comparison table of signal strength and distance or by estimating according to the fitting curve. Therefore, the application of this method is studied the most. However, this method also has its shortcomings. For example, drawing the fitting curve requires a lot of preparation and measurement work, and this method is sensitive to the environment. Changes in environmental factors will cause large positioning errors in the data model obtained in advance, so it must be re-measured, which increases the workload.

The above three ranging methods have their own advantages and disadvantages, and can be flexibly selected according to the required positioning accuracy in practical applications. When high accuracy is required, TDOA method is generally chosen, but this method requires high energy consumption and cost; When the accuracy requirement is not high, choose the RSSI method, which is simple to implement and does not require additional equipment.

After using the vehicle position real-time tracking algorithm proposed in this article to calculate the estimated value of the target node, it is used as the initial position of each target node for iterative optimization based on the initial value. The algorithm process is as follows:

(1) Initialize the population.

The estimated value obtained in the first step is used as the initial solution of the population to find the optimal solution. Based on the initial solution, the population is initialized using the uniform distribution function.

Assuming that the coordinate of the target node obtained by the weighted centroid algorithm based on RSSI is $A_0 (A_{0x}, A_{0y})$ in the first step, it is taken as the initial solution to initialize the population, and the initial value is added to improve the performance of the evolutionary algorithm, reduce the computing time and improve the efficiency.

Using the uniform distribution probability function, the population is initialized as:

$$\begin{aligned} X_{1i,G} &= rand[0, N(0, 1)(B - A)](B - A) + A_{0x} \\ X_{2i,G} &= rand[0, N(0, 1)(B - A)](B - A) + A_{0y} \end{aligned} \tag{8}$$

where, A and B are variable boundaries that utilize the distance between the target node and the anchor node; $N(0, 1)$ represent a standard Normal distribution.

(2) Mutation operation:

In numerous literature, through extensive experimental function testing, it has been found that different modes of differential evolution algorithms have certain differences in simulation performance. However, a large number of experiments have shown that Mode 1 and Mode 4 have the best performance, but Mode 1 is simpler compared. Therefore, in this algorithm, we chose the standard DE algorithm, which is the mutation mode using:

$$V_{i,G} = X_{1i,G} + F * (X_{1i,G} - X_{2i,G}) \tag{9}$$

In the equation, F represent the coefficient of variation.

(3) Cross operation:

The crossover operation is to increase the diversity of the newly generated population. According to the newly generated individual $V_{i,G+1}$, part of the code is exchanged between the old and new individuals according to the crossover strategy, so as to form a new individual.

Three different individuals r_i are randomly selected in the population, and a random index $R = (1, 2, \cdots, N)$ is selected. For each $i\ U(0, 1)$, the following operations are performed:

$$u_{ji,G+1} = \begin{cases} V_{ji,G+1}, j = i \\ X_{ji,G+1}, j \neq i \end{cases} \tag{10}$$

The formation of new individuals:

$$U_{i,G+1} = (u_{1i,G+1}, u_{2i,G+1}, \cdots, u_{ji,G+1})$$

In order to determine whether the newly generated individual $V_{i,G+1}$ will be left to participate in the evolution of the new generation $G + 1$ or eliminated, a selection operation is carried out.

Calculate the fitness function of $U_{ji,G+1}$ and $X_{ji,G+1}$ respectively, and compare their values according to the following formula to determine the survival of the fittest.

If $f(X_{ji,G+1}) < f(U_{ji,G+1})$, $X_{ji,G+1}$ will be eliminated, but instead $U_{ji,G+1}$ will enter the next generation.

(4) Set the fitness function to

$$f = \sqrt{(x_m - x)^2 + (y_m - y)^2} - d \tag{11}$$

In Formula (11), (x_m, y_m) is the coordinate value of the anchor node with the largest RSSI value, (x, y) is the vector involved in evolution, and d is the distance calculated according to the RSSI value.

In this way, after the completion of the set number of iterations, the output is the optimal solution, that is, the exact coordinates of the target node. Of course, the larger the number of iterations, the more accurate the value will be, but at the same time, it will increase the amount of calculation. Generally, 30~50 times of generation selection will be selected.

In summary, the real-time location tracking algorithm of wireless sensor network can more accurately calculate the real-time location information of intelligent logistics cold chain transport vehicles.

3 Experimental Analysis

3.1 Analysis of Real-Time Tracking Algorithms for Wireless Sensor Networks

For the establishment of target tracking simulation model, the nonlinear dynamic model is adopted. In this model, it is assumed that the target node only moves along the x-axis direction, then the dynamic null of this model can be expressed as:

$$x_{k+1} = \alpha x_k + \beta \frac{x_k}{1 + (x_k)^2} + \gamma \cos(1.2k) + w(k) \tag{12}$$

Simultaneously construct its observation equation as follows:

$$z_k = \frac{(x_{k+1})^2}{20} + v(k) \tag{13}$$

Among them, $x(k)$ represent the movement distance of target node k; $\gamma \cos(1.2k)$ are the loading terms of the system, $w(k)$ is the noise level of the state equation, and $v(k)$ is the noise level of the observation equation; α, β and γ represent parameter factors separately.

In the simulation experiment, the real-time tracking algorithm of wireless sensor networks is used for symmetric sampling strategy. Before testing, the effectiveness of the proposed method is verified and relevant experimental parameters are set as shown in Table 1.

Table 1. Parameter Settings

Parameter	Value
α	0.5
β	2
γ	8
Total time step N	100
$w(k)$ variance of noise level of Equation of state	4
The noise level $v(k)$ variance of the observation equation	0.04

Next, real-time tracking simulation will be conducted based on the above parameter settings. The tracking simulation results without parameter adjustment and after adjustment, as well as the tracking error, are shown in the following figure.

From Figs. 2 and 4, it can be seen that there is a significant difference between the tracked trajectory and the true trajectory of the target node before parameter adjustment. However, after appropriate parameter adjustments are made to the real-time tracking algorithm of the wireless sensor network, the filtering trajectory of the real-time tracking algorithm of the wireless sensor network is basically consistent with the true trajectory of the target node, achieving good tracking results.

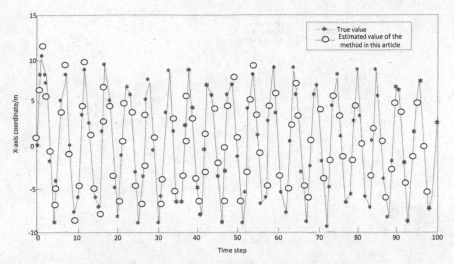

Fig. 2. Real time tracking simulation diagram

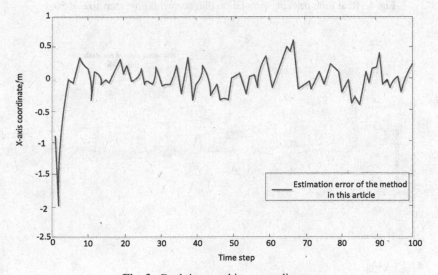

Fig. 3. Real time tracking error diagram

From Figs. 3 and 5, it can be seen that before parameter adjustment, the tracking prediction error fluctuates between [−2, 0.5], with a large fluctuation amplitude. However, after appropriate parameter adjustment of the real-time tracking algorithm for wireless sensor networks, the tracking prediction error fluctuates between [−0.5, 0.5], and the fluctuation fluctuation is small. Based on the above results, it can be concluded that the real-time tracking algorithm filtering algorithm based on deterministic symmetric sampling in wireless sensor networks has high tracking accuracy, can achieve efficient prediction and tracking in nonlinear models, and is effective.

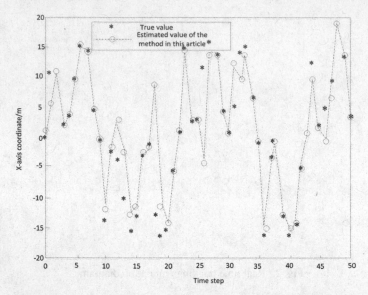

Fig. 4. Real time tracking simulation diagram with time step size = 50

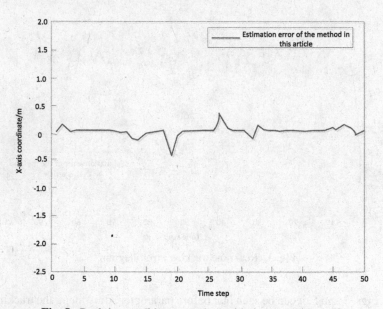

Fig. 5. Real time tracking error chart with time step size = 50

3.2 Comparison Simulation Experiment Between Real-Time Tracking Algorithm and EKF of Wireless Sensor Network

Next, to further verify the superiority of the proposed real-time tracking algorithm for wireless sensor networks, the real-time tracking algorithm for wireless sensor networks and the tracking performance of EKF on the same system model are analyzed. The results are shown in the following figure.

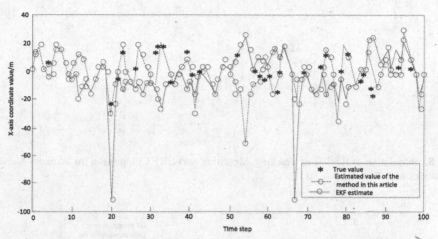

Fig. 6. Simulation of Real Time Tracking Algorithm and EKF Comparison for Wireless Sensor Networks with Time Step = 100

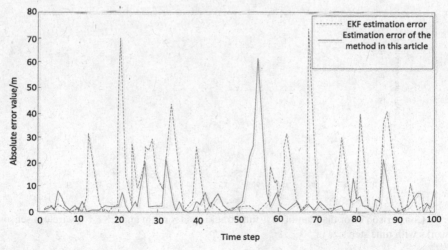

Fig. 7. Comparison error diagram of real-time tracking algorithm and EKF for wireless sensor networks with time step size of 100

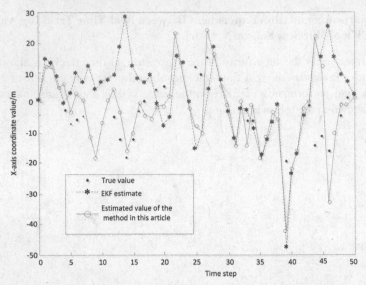

Fig. 8. Simulation of Real Time Tracking Algorithm and EKF Comparison for Wireless Sensor Networks with Time Step = 50

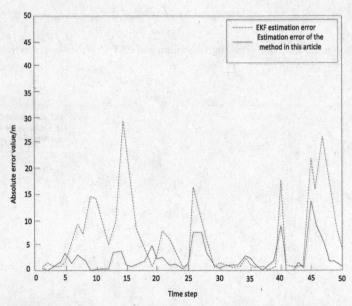

Fig. 9. Comparison error diagram of real-time tracking algorithm and EKF for wireless sensor networks with time step size of 50

Figure 6 and Fig. 8 respectively compare the tracking simulation effects of the system model with the time of 100 and 50. From the figure, it can be seen that in both cases, the real-time tracking algorithm of wireless sensor networks generally outperforms the

EKF algorithm in terms of tracking performance, and its tracking trajectory is closer to the actual trajectory, indicating that the real-time tracking algorithm of wireless sensor network adopts definite line sampling and UT transform to overcome the nonlinear transmission in EKF algorithm to a certain extent, and also proves that UT transform can describe the nonlinear characteristics of the system more accurately by using symmetric sampling strategy.

Figure 7 and Fig. 9 show the tracking errors for two scenarios with time steps of 100 and 50, respectively. It can be seen that there are errors in both the real-time tracking algorithm of wireless sensor networks and EKF in the tracking and prediction process, regardless of the scenario. However, compared to the two methods, the proposed method can effectively reduce tracking error, which can be controlled below 15. Although the EKF algorithm can also reduce the error, its error control results are still higher than the proposed method. Therefore, it indicates that the real-time tracking algorithm of wireless sensor networks has achieved more accurate predictions with smaller overall errors.

4 Conclusion

Although the real-time location tracking algorithm of intelligent logistics cold chain transport vehicle based on wireless sensor network proposed in this paper improves the accuracy of target tracking to a certain extent, it also has the following deficiencies:

In the real-time tracking algorithm of wireless sensor networks, due to the dependence on the initial solution proposed by coarse positioning, it is necessary to select the appropriate fitness function and parameters to make the algorithm converge quickly and find the optimal solution accurately. The next goal of real-time tracking algorithm of wireless sensor network is to use the existing limited resources, reduce the system energy consumption, improve the tracking accuracy and robustness of the algorithm. The research in three-dimensional space still needs to be strengthened. The location node tracking algorithm of intelligent logistics cold chain transport vehicles needs to be further improved to optimize the tracking efficiency.

References

1. Li-feng, W., Fei, H., Zhu, G.-H.: Design of cold chain logistics information real time tracking system based on wireless RFID technology. In: Liu, S., Ma, X. (eds.) ADHIP 2021. LNIC-SSITE, vol. 416, pp. 440–453. Springer, Cham (2022). https://doi.org/10.1007/978-3-030-94551-0_35
2. Gu, M.S., Miao, F., Gao, C.B., et al.: Research of localization algorithm of internet of vehicles based on intelligent transportation. In: Proceedings of the 2018 International Conference on Wavelet Analysis and Pattern Recognition (ICWAPR), pp. 056–061 (2018)
3. Li, G.: Development of cold chain logistics transportation system based on 5G network and Internet of things system. Microprocess. Microsyst., 021–035 (2020)
4. Meng, J.: Research on the early warning system of cold chain cargo based on OCR technology. World Eng. Technol. 003, 132–141 (2022)
5. Lian, J.: An optimization model of cross-docking scheduling of cold chain logistics based on fuzzy time window. J. Intell. Fuzzy Syst. Appl. Eng. Technol. (1), 41 (2021)

6. Fan, Y., Li, H.: Research on cold chain logistics operation mode under internet technology. J. Phys. Conf. Ser. **1972**(1), 012–039 (2021)
7. Chen, Z., Liu, X.C.: Statistical distance-based travel-time reliability measurement for freeway bottleneck identification and ranking. Transp. Res. Rec. **2675**(11), 424–438 (2021)
8. Zhang, X., Gou, H., Lv, Z., et al.: Double-quantitative distance measurement and classification learning based on the tri-level granular structure of neighborhood system. Knowl. Based Syst. **217**(6), 1–21 (2021)
9. Li, X., Wang, Z., Gao, S., et al.: An intelligent context-aware management framework for cold chain logistics distribution. IEEE Trans. Intell. Transp. Syst., 1–14 (2019)
10. Dalarmelina, N.D.V., Teixeira, M.A., Meneguette, R.I.: A real-time automatic plate recognition system based on optical character recognition and wireless sensor networks for ITS. Sensors **20**(1), 020 (2019)
11. Nnonyelu, C.J., Okide, C.P., Ahaneku, M.A.: Angle-of-arrival and time-of-arrival statistics of multipaths arising from scatterers on an elliptical disc with conic spatial density around the mobile. Phys. Commun. **49**(10), 10–18 (2021)
12. Gao, G., Liu, Y., Jinfeng, W.U., et al.: The real-time collection and monitoring system study of cold chain logistics information based on the internet of things. Heilongjiang Electr. Power, 005–013 (2018)
13. Alkayas, A., Chehadeh, M., Ayyad, A., et al.: Real-time identification and tuning of multirotors based on deep neural networks for accurate trajectory tracking under wind disturbances. 128–131 (2021)
14. Biradar, M., Mathapathi, B.: Security and energy aware clustering-based routing in wireless sensor network: hybrid nature-inspired algorithm for optimal cluster head selection. J. Interconnection Netw. **23**(1), 5–12 (2022)
15. Kim, J.: Hybrid TOA-DOA techniques for maneuvering underwater target tracking using the sensor nodes on the sea surface. Ocean Eng. **242**(15), 1–6 (2021)

A Real Time Tracking Method for Intelligent Logistics Delivery Based on Recurrent Neural Network

Xunyan Bao[1(✉)] and Dong'e Zhou[2]

[1] Zhejiang Changzheng Vocational and Technical College, Hangzhou 310012, China
baoxunyan@126.com
[2] Guangzhou Huashang Vocational College, Guangzhou 511300, China

Abstract. In order to further improve the real-time tracking effect of intelligent logistics distribution, this paper proposes a real-time tracking method for intelligent logistics distribution based on recurrent neural networks. Firstly, relevant analysis was conducted on real-time tracking of intelligent logistics delivery, and the planning process of order information and road information was determined. Secondly, considering the dynamic real-time traffic conditions and constantly updated customer orders analyzed above, an online target tracking model based on recurrent neural networks was established to predict the status of each node, and then correct the corresponding target status. Finally, Hungarian algorithm is used to solve the data association problem to reduce the detector error to the tracking algorithm. Finally, the Loss function is used to optimize the model performance and achieve accurate real-time tracking of intelligent logistics distribution. The results indicate the feasibility of the proposed method in practical applications, and by comparing it with other similar methods in solving the objective function, the advantages of this method in real-time tracking of intelligent logistics distribution are further verified, with high tracking accuracy.

Keywords: Smart logistics · Path planning · Real time tracking

1 Introduction

In recent years, with the continuous improvement of information technology and the rapid development of "online shopping era", the logistics industry has been an unprecedented spurt of development. Online shopping has many characteristics such as convenient transaction, low cost, rich profit and saving time, which makes the commercial institutions actively change their operation mode, while the real economy is also facing great challenges. After the online transaction is completed, the rest of the tasks are completed by the offline logistics. Therefore, the convenient service of e-commerce urgently needs a matching logistics system with low cost and high efficiency, so as to operate more stably [1, 2]. The compression of logistics costs and the diversified needs of consumers have become the bottleneck restricting the faster and more stable development of contemporary logistics. In this context, the concept of "intelligent logistics" came into being [3].

© ICST Institute for Computer Sciences, Social Informatics and Telecommunications Engineering 2024
Published by Springer Nature Switzerland AG 2024. All Rights Reserved
L. Yun et al. (Eds.): ADHIP 2023, LNICST 550, pp. 113–132, 2024.
https://doi.org/10.1007/978-3-031-50552-2_8

In recent years, China's logistics industry has undeniably achieved explosive break-throughs, and these changes in smart logistics are entirely attributed to the rapid pop-ularization and application of new generation advanced information technologies such as the global Internet of Things, cloud computing, and mobile internet in human life. Many advanced modern logistics facilities and equipment have advanced scientific and technological capabilities such as digitization, informatization, and intensification [4, 5]. Some logistics companies are increasingly using information technology in their logis-tics distribution business. However, there are still some problems that need to be solved. Firstly, due to the multiple links and participants involved in the process of logistics distribution, it is difficult to obtain and share information. Different logistics companies, suppliers and transporters use different tracking systems and data formats, resulting in information silos and data inconsistencies [6]. Second, existing tracking technologies also face challenges in complex urban environments. Factors such as tall buildings, traffic congestion, and signal interference may result in reduced positioning accuracy or inability to obtain accurate location information [7]. Therefore, in order to improve the visualization degree of logistics distribution, logistics companies and customers can understand the location and status of goods in real time, improve the efficiency and accu-racy of distribution, and carry out real-time tracking of intelligent logistics distribution is of great significance.

Therefore, this paper proposes a real-time tracking method of intelligent logistics distribution based on recurrent neural network to achieve accurate distribution tracking. This method firstly analyzes the real-time tracking of intelligent logistics distribution, and determines the planning process of two types of information, order information and road information. Secondly, considering the dynamic real-time traffic conditions analyzed above and the constantly updated customer orders and other factors, an online target tracking model based on recursive neural network is established to predict the status of each node. Then modify the corresponding target state. Finally, the Hungarian algorithm is used to solve the data association problem to reduce the detector error to the tracking algorithm. Finally, the loss function is used to optimize the model performance and achieve accurate real-time tracking of intelligent logistics distribution.

2   Real Time Tracking Analysis of Intelligent Logistics Delivery

Real-time dynamic path planning of intelligent logistics refers to that when the real-time information that has a great influence on the path selection is conveyed to the system center, the system needs to make targeted responses to obtain the optimal path of the current situation. Real-time information here includes new order placing, order cancellation, order delay, road speed change, road construction, traffic congestion, etc., which can be roughly divided into two categories: order information and road information [8]. Although the real-time dynamic path planning system is said to be a re-optimization strategy, it does not completely re-plan from the beginning. But during the driving process, based on the real-time information received, new unserved customer point paths are added to the existing path planning to complete the overall path re planning [9]. Below is a detailed introduction to the process of re planning based on two categories of information: order information and road information:

(1) When the road information changes, the system will first automatically detect whether the planned delivery path passes through that road. If the system automatically captures the road speed RS (Road Speed) of the current driving section, then finds the previous customer point i_p that has completed the service, records its position j, and calculates the arrival time of j to all unassigned points. The point with the shortest time is assigned as the key point i_1, and gradually continues to rank the remaining customer points in descending order of time as the key points for the next service goal, Then, for each key point recorded, the system automatically removes it from the set of unassigned points until all final orders are re optimized; If not, the system will automatically ignore this information [10]. The reason for this arrangement is that the path before this customer point is already in the optimal state adjusted by the system, so there is no need to optimize it again. This not only saves computational time for the system, but also simplifies the complexity of model calculations. The specific calculation process is shown in Fig. 1:

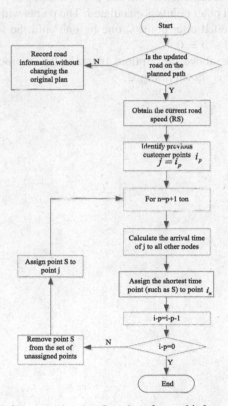

Fig. 1. Real time path planning flowchart for road information changes

(2) When the order information is changed, it is similar to the road information change strategy mentioned above. When A new order i_p arrives in the system, the first step is to determine whether the order is on the planned path. If yes, insert it between the customer points before and after the original path; If not, first calculate the distance d_{pn} between the closest customer points i_1 and i_p to the central system, and then calculate the distance $d_{n,n-1}$ between i_n and the previous customer point i_{n-1}, and then compare d_{pn} and $d_{n,n-1}$.

① If $d_{pn} > d_{n,n-1}$, then the new path is $i_{n-1} - i_p - i_n$, that is, the new order is inserted between i_{n-1} and i_n. The system has been optimized before i_{n-1} and after i_n, so it remains unchanged.

② If $d_{n,n-1} > d_{pn}$, the new path is $i_{n-1} - i_n - i_p$, that is, new orders are placed behind these two points. In this way, according to the non after effect of dynamic programming, the system needs to resort and optimize all customer points after the new order i_p.

Then, point i_{n-1} is regarded as the key point, marked as j, and the arrival time from point i_{n-1} to all other points is calculated. The points with the minimum value are selected and inserted after point i_n one by one until the set of all unallocated points is 0, namely, $i - p = 0$.

According to the above description, the calculation process when the order information changes is shown in Fig. 2:

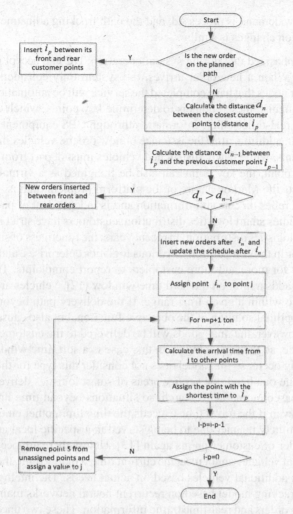

Fig. 2. Real-time path planning flow chart of order information change

(3) An example of intelligent logistics distribution in a certain region is introduced here
to demonstrate the characteristics of real-time dynamic path tracking. As shown in
Fig. 3, the distribution center (DC) has 3 vehicles and 9 customer points in its initial
state (a), and two vehicles are dispatched by the distribution center to complete the
distribution task. The distribution scheme of the first car is DC-b-a-h-e-f-DC, and
the distribution scheme of the second car is DC-d-o-g-c-DC (b). At some point, the
distribution center received four new orders, which are q, k, l and r (c). Then the
distribution center will readjust the route according to the current key points and real-
time road traffic information. Figure (c) shows that the key points set $N_c(\varepsilon) = (b, o)$,
the unallocated points set $N_U(\varepsilon) = (a, h, e, f, g, c, o)$, and the new demand points
q, k, l and r. After route replanning, the distribution route of the first vehicle is DC-
b-a-q-h-e-k-DC, and the route of the second vehicle is DC-o-l-g-m-c-f-DC (d). In

this way, the new demand is processed, and the path tracking adjustment caused by traffic information changes is similar.

It should be emphasized that during system operation, DC can accept new demand orders at any time. When a new order arrives, the system only considers unassigned nodes, and customer nodes that have completed the service will be automatically deleted. The method to distinguish these points is to determine key points, which can be determined by capturing real-time vehicle information through GPS equipment in the distribution center [11]. In addition, after the issuance of new orders, vehicles in transit must depart from key points, while newly dispatched vehicles must depart from the distribution center. At this point, the key point can also be imagined as a virtual distribution center, which reflects the Markov nature of the platform problem, that is, the planning of future paths only relies on current information and is independent of past solutions.

As this paper studies smart logistics distribution, customers have strict requirements on the timeliness of logistics distribution. In recent years, the timeliness of smart logistics is also a major factor in the competition of various logistics enterprises, and has become an important aspect for more and more customers to report complaints. Therefore, the model in this paper adds a limit on the hard time window [12]. Vehicles are required to complete the service within a given time range. If the delivery path beyond this range is not only not the optimal solution of the objective function, but also causes the loss of waiting or delay. However, the final goods will be delivered to the customer beyond the time limit in advance, so the time window in this case is a soft time window problem. This does not usually occur, so the model does not consider this type for the time being.

In real life, on the one hand, the service areas of smart logistics delivery are determined in advance based on market research, so situations beyond time limits are rare; On the other hand, even if the travel time exceeds the time limit, other emergency vehicles can arrive in a timely manner due to being served in a specific local area, and only need to plan the order of customer points again [13]. Unless there are occasional large orders and the current vehicle cannot meet the current time window limit, the distribution center will dispatch additional vehicles based on actual needs. The intelligent logistics delivery real-time tracking model based on recurrent neural networks mainly considers two factors: random orders and real-time traffic information. These two uncertain factors can make it more difficult to grasp the delivery time of delivery vehicles, so the calculation process is slightly more complex than the static platform problem [14]. The system dispatch center must provide a pruning plan in a short period of time after receiving real-time information, with strict timeliness requirements, which increasingly require high calculation speed and accurate and appropriate results. Therefore, the next step is to conduct research on the intelligent logistics delivery real-time tracking model based on recurrent neural networks, in order to achieve precise intelligent logistics delivery real-time tracking and improve service response speed.

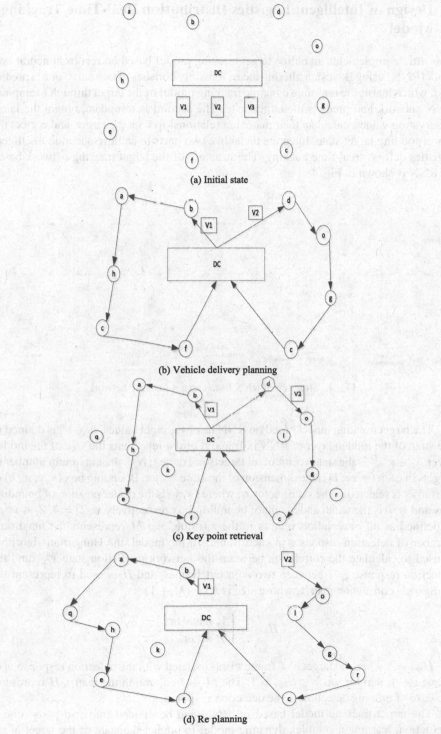

(a) Initial state

(b) Vehicle delivery planning

(c) Key point retrieval

(d) Re planning

Fig. 3. Schematic diagram of real-time path re planning

3   Design of Intelligent Logistics Distribution Real-Time Tracking Model

This article implements an online target tracking model based on recurrent neural network (RNN) using Bayesian filtering ideas. It mainly consists of two parts: one is prediction, which learns the real-time dynamic tracking model of the target through a temporal RNN network and predicts the target state; The second is to update, obtain the latest observation values, calculate their matching relationship with the target, and correct the corresponding target state. Integrate the above two parts to achieve accurate intelligent logistics delivery real-time tracking. The structure of the target tracking network based on RNN is shown in Fig. 4.

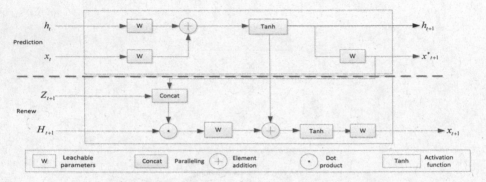

Fig. 4.   Structure of RNN based target tracking network

The target tracking model based on RNN has four input values. $h_t \in v^n$ is defined as the state of the hidden layer of RNN in frame t, and n represents the size of the hidden layer. $x_t \in v^{N,D}$ is the state vector of all targets in frame t; N is the maximum number of targets in each frame; D is the dimension of the state vector. Bounding box (x, y, w, h) of the target is selected as the state vector x_t, where (x, y) is the center position of bounding box and w, h is the width and height of bounding box respectively, so $D = 4$. $Z_t \in v^{M,D}$ is defined as all observation vectors in the t frame, and M represents the maximum number of detection responses in each frame. In the model, the Hungarian algorithm is used to calculate the correlation between the network prediction state x_{t+1}^* and the detection response z_{t+1} between two adjacent frames, and H is used to represent the Hungarian correlation matrix, whose size is $N \times (M + 1)$

$$H_{nm} = \begin{cases} 1, & n \text{ and } m \\ 0, & \text{other} \end{cases} \tag{1}$$

$H_{nm} = 1$ when the target n of frame t is associated with the detection response m of frame $t + 1$, and for $\forall n$, $\sum_m H_{nm} = 1$. The $M + 1$ column in the matrix H represents the case of missing detection in the detection response.

The target tracking model based on RNN can be divided into two parts: one is prediction, learning a complex dynamic model to predict the state of the target at the next moment; The second is updating, measuring the matching relationship between the target and the observed value of the next frame, and correcting the target state according to the corresponding observed value.

(1) Prediction stage. Assuming that the initial state x_0 of the target and the initial state h_0 of the network hidden layer are known. Firstly, the RNN network model is used to predict the possible state of the target. At time t, given the state h_t of the hidden layer and the target state x_t, the hidden layer state at time $t + 1$ can be obtained

$$h_{t+1}^l = \tanh\left(W_h h_t^l + W_x x_t\right) \tag{2}$$

where, W_h and W_x represent learnable parameters in the full connection layer.
From this, it can get the predicted state of the target

$$x_{t+1}^* = W_{x*} \tanh\left(W_h h_t^l + W_x x_t\right) \tag{3}$$

From the above equation, it can be seen that the predicted value x_{t+1}^* of the target state at time $t + 1$ only depends on the hidden layer state h_t and target state x_t at time t. That is to say, the RNN model only predicts the possible state of the target at the next moment based on its current state.

(2) Update stage. After the latest observed value z_{t+1} is obtained, the predicted value of the network can be updated. It is noted that when tracking multiple targets, multiple detection responses will be obtained in each frame. At this time, it is necessary to determine which detection response should update the corresponding target, which is a data association problem. In this paper, the Hungarian algorithm is used to solve this problem [15]. By constructing the cost matrix, the solution matrix that minimizes the correlation cost between data is found, so as to obtain the optimal allocation.

At moment t, the target state x_t is known and the latest observation value z_{t+1} is obtained. The euclidene distance d between the predicted location (x_n, y_n) of each target bounding box and the bounding box position (x_m, y_m) of all observations in the prediction stage is calculated successively.

$$d_{nm} = \sqrt{(x_n - x_m)^2 - (y_n - y_m)^2} \tag{4}$$

Build a cost matrix C with size $N \times (M + 1)$, the expression is as follows:

$$C = \begin{bmatrix} d_{11} & d_{12} & \cdots & d_{1j} & \sigma \\ d_{21} & d_{22} & \cdots & d_{2j} & \sigma \\ \cdots & \cdots & \cdots & \cdots & \sigma \\ d_{i1} & d_{i2} & \cdots & d_{ij} & \sigma \end{bmatrix} \tag{5}$$

where d_{nm} represents the distance measurement between target n and all detection responses m, the number of targets is N, and the number of detection responses is M. The distance threshold σ is artificially set. If the distance between the target n and all detection responses is greater than σ, it is determined that the target n is missed and column $M + 1$ in the solution matrix H is marked as 1.

With the cost matrix C, the matching relation between the target and the detection response can be obtained by finding the solution matrix H_{t+1} which minimizes the correlation cost through the Hungarian algorithm. If column m of row n in H_{t+1} is marked as 1, it means that the m detection response z_{t+1}^m is assigned to the n target x_t^n, and after obtaining H_{t+1}, the target state x_t^n can be updated by the corresponding observed value z_{t+1}^m according to the matching relation.

Then connect the predicted value O of the network in parallel with the observed value P, denoted as Q. The update of the target state at time R can be calculated as follows:

$$x_{t+1} = W_{x_{t+1}} \tanh\left(h_{t+1} + W_H\left(H_{t+1} \bullet z_{t+1}^*\right)\right) \tag{6}$$

Next, let's do the dot product between H_{t+1} and z_{t+1}^*. The state update in the model relies on the predicted state x_{t+1}^* at the current time, the latest observation vector z_{t+1}, and the matching relationship H_{t+1} between the state and observation to correct the target's state.

Because the association algorithm is added, the algorithm itself has a certain robustness to the error caused by the detector. In case of missing detection, the undetected target n will be marked as 1 in the n row and $M + 1$ column of the incidence matrix H, that is, it will be judged as invalid detection, and the target state will not be updated, so as to minimize the impact of the detector error on the tracking algorithm.

After constructing the tracking model, we need to calculate the optimal model through the loss function, which measures the error between the predicted value and the real solution. The model performance can be optimized by minimizing the loss. For tracking algorithms, loss functions are defined mainly by measuring their tracking performance. However, the emphasis of tracking performance evaluation varies with different application scenarios. For example, in football matches, we tend to pay more attention to the correct identification of athletes and try to avoid the situation of identity ID conversion, while in vehicle assistance systems, tracking algorithms are required to have a high accuracy and recall rate to avoid accidents. For this algorithm, we hope that it can have a good performance in tracking accuracy, so the loss is defined as follows:

$$L\left(\tilde{x}, x^*, x\right) = \frac{\lambda}{ND} \sum \left\|x^* - \tilde{x}\right\|^2 + \frac{\kappa}{ND} \sum \left\|x - \tilde{x}\right\|^2 \tag{7}$$

Among them, $\tilde{x}$ represents the true value, x^* and x represent the predicted and updated values, N represents the number of targets, D represents the feature dimension, and λ and κ are constants. Note that the loss here is the sum of the losses of the training samples in all frames of the video sequence. From the definition of loss, it can be seen that we hope the trajectory predicted by the network can be as close to the real trajectory as possible. Therefore, the model is optimized by minimizing the Mean squared error between the predicted value and the updated value of the target state and the actual value to improve the tracking accuracy.

4 Experimental Analysis

4.1 Example Parameter Setting and Evaluation Index

In this model, weight setting and parameter setting are two important factors for solving the optimal solution of the objective function. Their values directly affect the accuracy of the model's calculation results, and thus affect the overall efficiency of real-time tracking of intelligent logistics distribution. So in order to avoid errors in arbitrary guesses, this article provides the following reference data:

(1) The weight of delivery time for vehicles in transit $\alpha = 0.6$
(2) Weight of waiting time before service $\beta = 0.2$
(3) The new order must be completed within two hours after the sub acceptance is issued; The initial order is completed within one and a half hours after the vehicle leaves the distribution center.
(4) The weight of the waiting time between the completion of the service and the time when the vehicle leaves the customer point is $\gamma = 0.1$
(5) Random number weight $\theta = 1.3$ for new orders.
(6) The conditions for determining the deployment of additional vehicles: unserved customer points exceed 0.5% of the total number.
(7) Set the interval for collecting real-time information to 10 min.
(8) Jitter process:

$$P_{insert} = 0.3 \quad P_{cross} = 0.2 \quad P_{incross} = 0.15$$

Domain search operator probability: $P_{or-opt} = 0.5 \quad P_{2-opt} = 1 - P_{or-opt} = 0.5$.

In order to improve the timeliness of the smart logistics distribution real-time tracking model, this paper defaults that the orders of the system come from the initial time of the planning period, and the other 50% of the orders are received by the distribution center later, randomly generated during the operation of the system. The random number determination method is as follows: $\max(0, S_i - \theta d_{oi} - r)$.

The maximum payload of each vehicle is 7.5 tons, and the maximum distance for each delivery must be within 30 km. According to the company's order system, we have obtained the weight of the goods required for each customer point and the vertical distance from the distribution center, as shown in Table 1:

Table 1. Customer point distance and demand statistics

Information Customer points	Cargo weight kg	Distance km
8	2	6.5
15	4.5	10
19	2.6	18
34	4	27
36	5.1	21

In this paper, the advantages and disadvantages of the calculation method considering the platform problem are mainly based on: the length of distribution time, the total time used to adjust the scheme, the number of cars dispatched, the waiting time of customers and the loss statistics caused by the number of orders refused. But the objective function only considers the timeliness of the scheme from the perspective of time.

4.2 Path Planning for Real-Time Tracking of Intelligent Logistics Distribution

According to the Road Traffic Management Bureau, based on the calculation software of highway operating speed, after inputting a street in a certain area into an electronic scanning measuring instrument, the software measures the average speed and real-time vehicle speed of each road segment in the distribution of all customer points in the area. The specific test results are shown in Table 2:

Table 2. Real time road speed statistics

Road type	Average speed	Vehicle speed	Covering road sections
Primary road	60 km/h	45 km/h	22⇔23, 22⇔34, 24⇔25
Secondary road	40 km/h	30 km/h	1⇔2, 2⇔3, 1⇔7, 7⇔9
Third level road	30 km/h	20 km/h	All others

The initial position of the vehicle is at 1:00. According to the company's order system, the current customer orders requiring service are: 8, 15, 19, 34, 36. According to the actual road conditions, we stipulate that direction 34 → 37 is a one-way street. According to the shortest path method, the nodes of the initial path are sorted as 1–8–19–15–34–36. The delivery time and shortest path data of each customer point provided according to the equipped on-board recorder are shown in Table 3: Time (min).

Table 3. Statistics of Shortest Distance Nodes and Shortest Delivery Times for Real Time Tracking of Intelligent Logistics Delivery Paths

Node	1-8	8-19	19-15	15-34	34-36
Route	1-7-8	8-12-19	19-16-15	15-24-23-22-34	34-35-36
Delivery time	1.24	1.85	1.47	2.91	1.63

According to the table above, the shortest initial path is:
1-7-8-12-19-16-15-24-23-22-34-35-36.
The expected shortest delivery time is $1.24 + 185 + 1.47 + 2.91 + 163 = 9.1$ min.
The specific real-time tracking route is shown in the figure (Fig. 5):

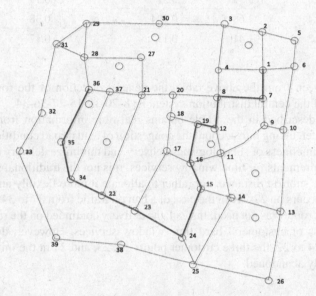

Fig. 5. Path planning diagram for real-time tracking of smart logistics delivery

However, when the distribution system started to operate, the real-time information collected by the central system for the first time was displayed 10 min later, and the new order information and road condition information were received after 8:00. Then 8 is the key point. All nodes before 8 o'clock and 8 o'clock are not considered in the replanning. Two new orders appear at 20 o'clock and 27 o'clock. 24 → 22 direction due to rush in the rush hour, the road traffic has sustained congestion, so that the traffic is severely paralyzed, all vehicles can not walk, at this time the traffic speed on the road is almost zero.

4.3 Algorithm Analysis

According to real-time information and initial order (except 8), the next six customer points to be served are 15, 19, 20, 27, 34 and 36.

The program based on recursive neural network algorithm was programmed into the MATLAB platform, all parameters were set, the recorded data collected in the input stage was clicked to start calculation, and the output calculation results were shown in Table 4: (Time: min).

Table 4. Adjusted inter node delivery time statistics for road sections

Road section	Target value	Delivery time	Wait time before service	Wait time after service
8 → 20	1.9	2.1	0.3	0.24
20 → 19	2.4	1.7	0.18	0.245
19 → 15	3.8	2.03	0.28	0.125
15 → 27	4.8	5.34	0.31	0.19
27 → 36	3.4	2.68	0.194	0.38
36 → 34	3.2	1.41	0.11	0.17

As can be seen from the above table, the general direction of the route after the redistribution of the central distribution system is: 8-20-19-15-27-36-34.

The model described in the article obtains real-time information from the traffic information center, taking into account the congestion of actual road conditions. In order to improve the timeliness of smart logistics delivery and minimize delivery time to meet customers' requirements for time window services, it is not the traditional solution that aims to have the shortest distance, but rather to allocate it more flexibly and randomly. So congestion occurs on 27-21; In the case of a ban on traffic from 37 to 34, the optimal lane for the next route was not used. Instead, the delivery continued on the road, catering to the timeliness of customers' hard time window services. However, due to severe congestion on 24 to 22, the three customer points 22, 23, and 24 in the initial path had to be temporarily abandoned.

Table 5. Statistics of objective function and actual value after improvement scheme

Road section	Target value	Actual value	Relative drop
8 → 20	1.9	2.64	−0.74
20 → 19	2.4	2.125	0.275
19 → 15	3.8	2.435	1.365
15 → 27	4.8	5.84	−1.04
27 → 36	3.4	3.254	0.146
36 → 34	3.2	4.89	−1.69

According to the statistical results in Table 5, it can be seen that the actual running time value of the distribution route after adjustment is still lower than the target value in the initial plan. However, due to the congestion of individual sections, the lag of updating information, and the response buffer time of receiving new orders, the overall service time of three sections is still lower than the target value.

Figure 6 shows the planned paths of all nodes after system readjustment.

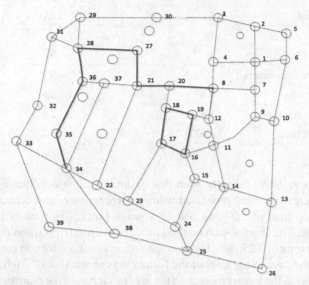

Fig. 6. Path planning scheme diagram based on real-time tracking information

It should be noted that in the figure, a double line indicates that the driving vehicle has passed through the road section twice, but the direction is opposite. Therefore, according to the above figure, the adjusted final vehicle planning path scheme is: 8-20-18-19-16-15-16-17-18-20-21-21-28-28-36-35-34.

The final delivery time used is $2.1 + 1.7 + 2.03 + 5.34 + 1.68 + 1.41 = 14.26$ min.

On the basis of the above case study, this article conducted an experiment to increase the number of regional customer points of M company in a certain location to 100. In order to prove the feasibility of the algorithm and model, and expand the research scope, a relatively complex platform problem study was conducted for single yard and multi vehicle models. The shortest path method and recursive neural network algorithm were used to calculate 6 times, and the average value was taken using MATLAB, The maximum number of iterations for the search algorithm set in the program is 5. The final result analysis is shown in Table 6: (Time: min).

Among them, the initial path creation time refers to the time taken by the two algorithms in the first stage of the planned path. Since both algorithms use Dijkstra method, the time used is the same. The running time of path optimization refers to the total time after the planned path is changed according to dynamic real-time information until the delivery plan is completed. The last column of path readjustment time refers to the sum of the first two columns, that is, the total time of the whole scheme implementation.

Table 6. Results analysis of 100 customer points processed based on recursive neural network algorithm and Dijkstra method

Project	Recurrent neural networks	Shortest path method
Target value	1.789	2.635
Aggregate demand	84	75
New demand	14	8
Number of vehicles	4	6
Number of rejected customers	2	0
Initial path build time	0.67	0.67
Path optimization run time	1.4	2.6
Path readjustment time	2.07	3.27

From the above table, it can be seen that under the same conditions, a total of 98 requirements were effectively processed using the recursive neural network algorithm, with a calculation time of 2.07 min and an average processing time of 0.021 min for each requirement; The shortest path method processed a total of 83 requirements, with a total calculation time of 3.27 min. The average processing time for one requirement was 0.039 min, and the processing level of the former was about 46.15% higher than that of the latter. Not only is the former more than the latter in terms of the number of processing requirements during the same period, but the former is far less than the latter in terms of the number of vehicles used. In addition, the former considers that due to real-time information, the calculation results are more realistic, thus having greater advantages in dealing with practical problems. When dealing with large-scale implementation path optimization problems, algorithm runtime is an important consideration. This article has also done a lot of work in this area, adopting the 2-opt positioning operator and repeated positioning operator to synchronize, and in the local search stage, a portion of customer points are first located. This greatly promotes convergence speed, and the two operators synchronize the search domain structure, successfully increasing the accuracy of the results.

It can be seen from the previous introduction that the variable domain search algorithm is essentially an iterative process, so the reasonable design of convergence criteria is also a key factor to determine the computational workload. However, the setting of the number of iterations determines the amount of computation and further affects the time of scheme adjustment, so the reasonable setting of the number of iterations is also very important. In this paper, the number of iterations is set as 5. In order to verify the rationality of such setting, the sensitivity analysis of the iterations of the target value is also made, as shown in Fig. 7: Time (s).

According to the sensitivity analysis results, it can be seen that after the function is calculated 4 times, the target value gradually stabilizes, so it is reasonable to choose 5 iterations in the article.

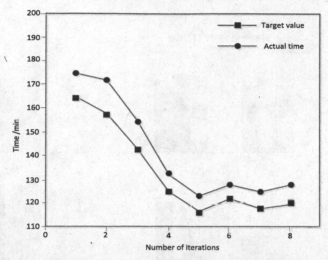

Fig. 7. Sensitivity Analysis of Target Value - Iteration Calculation Times

In order to prove that the optimized recursive neural network algorithm has more advantages than the Tabu search method and ant colony algorithm in solving such platforms under the same conditions, this paper, on the basis of unchanged previous parameter settings, conducts experiments respectively on 39 customer point cases, 60 customer points and 120 customer points in this area. The system takes the average value through eight calculations. The final comparison results are shown in Table 7: (Time: min).

Table 7. Comparison analysis of calculation results between recursive neural network algorithm and other methods

	RNN algorithm		Tabu search method		Ant colony algorithm	
	Average time	Use a car	Average time	Use a car	Average time	Use a car
39	1.426	2	1.26	3	1.41	3
60	1.624	3	1.65	5	1.503	5
120	2.29	5	2.45	6	2.06	6

Three methods can be calculated to handle the average order of one car: 2117, 15, 15. The average time of each order processed by the four methods is 1.78 min, 1.80 min and 1.85 min respectively. Obviously, it can be seen that the recursive neural network algorithm is the most efficient algorithm.

In order to more intuitively compare the advantages of recursion-based neural network algorithm compared with other algorithms of the same kind, origin software is used in this paper to make a drawing, and a simple analysis is shown in Fig. 8:

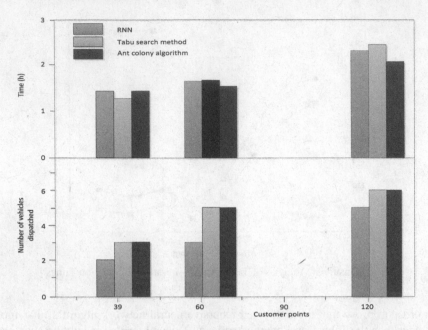

Fig. 8. Comparison of three algorithms

As can be seen from the figure above, under the same conditions, the algorithm based on recursive neural network has obvious advantages in the average time and the number of vehicles, but compared with tabu search method and ant colony algorithm, the calculation time is slightly longer than the latter two, but the number of vehicles used is less than the latter two. These differences are caused by the constant occurrence of uncertain information, which is inevitable in real life. However, considering all the factors, the algorithm based on recursive neural network is an ideal choice. In fact, when processing orders of less than 100 customer points based on recursive neural network algorithm, the response speed is lower than 0.5min, but due to the hysteresis of the actual operation, feedback and information transmission of various parts of the system, the calculation result has to be lower than the expected level. Therefore, with the increasing popularity of smart logistics in the future, this problem will soon be improved.

In order to further verify the advantages of the proposed method, the real-time tracking effects of RNN algorithm, Tabu search algorithm and Ant colony optimization algorithms were tested respectively. 500 nodes were tracked for distribution, and their tracking accuracy was counted. The comparison results are shown in Fig. 9 below.

According to the results obtained in Fig. 9, it can be seen that the tracking accuracy of the proposed algorithm does not show a decreasing trend with the increase of nodes, always maintaining around 99%. However, the real-time tracking accuracy of the other two algorithms shows a decreasing trend with the increase of nodes, and when the delivery real-time tracking node reaches 500, its tracking accuracy is 79% and 74%, respectively. Compared with the three algorithms, the proposed algorithm has better real-time tracking performance and more practical applicability.

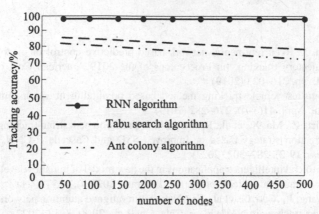

Fig. 9. Comparison Results of Tracking Accuracy

5 Conclusion

The proposal of intelligent logistics conforms to the new trend of the development of logistics industry in various countries, and modern logistics enterprises basically have various cutting-edge technologies of information and intelligence. Intelligent logistics distribution route based on real-time information is a hot topic in many academic circles. Therefore, in order to solve the problem of poor real-time tracking effect, this paper proposes a real-time tracking method of intelligent logistics distribution based on recurrent neural network. Firstly, the real-time tracking of intelligent logistics distribution is analyzed, and the planning process of two types of information, order information and road information, is determined. Secondly, considering the dynamic real-time traffic conditions analyzed above and the constantly updated customer orders and other factors, an online target tracking model based on recursive neural network is established to predict the state of each node, and then correct the corresponding target state. Finally, the Hungarian algorithm is used to solve the data association problem to reduce the detector error to the tracking algorithm. Finally, the loss function is used to optimize the model performance and achieve accurate real-time tracking of intelligent logistics distribution. The results show that the proposed method is feasible in practical application, and by comparing with other similar methods in solving the objective function, the advantages of the proposed method in real-time tracking of intelligent logistics distribution are further verified, and it has high tracking accuracy. In the future, how to design more efficient jitter methods and operators is a direction of our future research. In addition, how to quickly obtain real-time information in a short period of time, and match it with the real road conditions and electronic maps, and then quickly process and complete the planning of new schemes is also a problem that we need to continue to in-depth study. Finally, more accurate, faster and more comprehensive optimization is needed in the research of the solution algorithm.

References

1. Xie, S., Ren, J.: Recurrent-neural-network-based predictive control of piezo actuators for precision trajectory tracking. In: Proceedings of the 2019 American Control Conference (ACC). IEEE, pp. 016–023 (2019)
2. Hu, M.: Logistics vehicle tracking method based on intelligent vision. Int. J. Comput. Appl.Comput. Appl. **41**(3–4), 276–282 (2019)
3. Yang, S., Chen, Z., Ma, X., et al.: Real-time high-precision pedestrian tracking: a detection-tracking-correction strategy based on improved SSD and Cascade R-CNN. J. Real-Time Image Process. **19**(2), 287–302 (2022)
4. Wang, S.: Artificial intelligence applications in the new model of logistics development based on wireless communication technology. Sci. Program. **2021**(9), 1–5 (2021)
5. Liang, Z., Wang, J., Xiao, G., et al.: FAANet: feature-aligned attention network for real-time multiple object tracking in UAV videos. Chin. Opt. Lett. **20**(8), 8–17 (2022)
6. Voigt, S., Kuhn, H.: Crowdsourced logistics: The pickup and delivery problem with transshipments and occasional drivers. Networks **79**(3), 403–426 (2021)
7. Wang, Y., Peng, S., Guan, X., et al.: Collaborative logistics pickup and delivery problem with eco-packages based on time-space network. Expert Syst. Appl. **170**(3), 1–24 (2021)
8. Han, Q.H.: Research on the construction of cold chain logistics intelligent system based on 5G ubiquitous internet of things. J. Sens. **11**(6), 1–11 (2021)
9. Ma, L., Zhang, Y., Du, Y., et al.: Research on the framework of full-process condition monitoring and evaluation method for express logistics based on multi-information fusion and intelligent identification. IOP Conf. Ser. Mater. Sci. Eng. **740**(1), 8–19 (2020)
10. Chen, Y.T., Sun, E.W., Chang, M.F., et al.: Pragmatic real-time logistics management with traffic IoT infrastructure: Big data predictive analytics of freight travel time for Logistics 4.0. Int. J. Prod. Econ. **238**(8), 1–27 (2021)
11. Gao, D.: Design and development of intelligent logistics tracking system based on computer algorithm. J. Phys. Conf. Ser. **2074**(1), 2–12 (2021)
12. Feng, W., Wu, Y., Fan, Y.: A new method for the prediction of network security situations based on recurrent neural network with gated recurrent unit. Int. J. Intell. Comput. Cybern. **13**(1), 25–39 (2020)
13. Teng, S.: Route planning method for cross-border e-commerce logistics of agricultural products based on recurrent neural network. Soft. Comput.Comput. **15**(18), 12107–12116 (2021)
14. Xu, J., Wang, K., Lin, C., et al.: FM-GRU: a time series prediction method for water quality based on seq2seq framework. Water **13**(8), 1031 (2021)
15. Gan, H., Ou, M., Zhao, F., et al.: Automated piglet tracking using a single convolutional neural network. Biosyst. Eng.. Eng. **205**(1), 48–63 (2021)

A Real Time Tracking Method for Unmanned Traffic Vehicle Paths Based on Electronic Tags

Yarong Zhou and Bing Yuan(✉)

Faculty of Management, Chongqing College of Architecture and Technology,
Chongqing 401331, China
15309910368@189.cn

Abstract. In order to improve the real time tracking accuracy of the path and reduce tracking bias, a real-time path tracking method for unmanned vehicles based on electronic tags is proposed. Firstly, electronic tags are used to collect the path information of non human traffic vehicles. Secondly, based on the collected information, an improved Pure Pursuit algorithm is used to calculate the curvature of the pedestrian free flow vehicle path. Finally, construct an objective function for real-time path tracking, and solve the objective function to achieve real-time tracking of unmanned vehicle paths. The experimental results show that compared with existing tracking methods, the real-time tracking accuracy of the proposed method is consistent with the actual path, and the tracking deviation is significantly reduced.

Keywords: Electronic Labels · No Character Streaming Car · Real Time Path Tracking

1 Introduction

The unmanned factory is an important embodiment of the fact of intelligent factory. The unmanned factory logistics is the key technology to realize the unmanned factory, and the vehicle path tracking technology is one of the core technologies of the unmanned logistics vehicle driving system [1]. Vehicle autonomous driving is a hot research field at present, aiming at the special scenes of enclosed factories and the point-to-point vehicle autonomous driving that can be completely unmanned, such as the unmanned logistics system of factory materials transportation [2, 3]. The realization of unmanned logistics in the factory area can improve the automation and intelligence of the factory area, ensure the working environment of employees, improve logistics efficiency and reduce the cost of logistics in the factory area. The unmanned logistics vehicle mainly adopts the vehicle with specific driving device as the traction vehicle, and uses the support hook device to mount multiple material trailers behind the tractor. In order to achieve the purpose of flexible steering. Compared with ordinary passenger cars, unmanned logistics tractor is more sensitive to the change of front wheel deflection Angle due to the short wheelbase of front and rear wheels, and has higher requirements on the accuracy of vehicle lateral

© ICST Institute for Computer Sciences, Social Informatics and Telecommunications Engineering 2024
Published by Springer Nature Switzerland AG 2024. All Rights Reserved
L. Yun et al. (Eds.): ADHIP 2023, LNICST 550, pp. 133–149, 2024.
https://doi.org/10.1007/978-3-031-50552-2_9

control and control model. Path tracking control is one of the core technologies for the implementation of unmanned logistics vehicles, that is, by obtaining the real-time trajectory planned by the vehicle in real time, the effective path tracking algorithm is used to realize the control of the vehicle, so as to achieve the purpose of the vehicle driving on the correct route.

Reference [4] proposes a real-time path tracking method for unmanned vehicles based on the improved Stanley algorithm. Based on the characteristics of the vehicle's heading angle, lateral deviation, vehicle speed, and appropriate preview distance, a pure tracking algorithm is used to improve the calculation method of wheel angle in the Stanley algorithm. A new fusion algorithm is proposed to calculate the appropriate wheel angle of the vehicle at the current speed in real-time. Reference [5] proposed a real-time path tracking method for vehicles without people based on fuzzy Sliding mode control. First, the vehicle model was established based on the dynamic characteristics of two degrees of freedom; Secondly, in order to reduce the chattering problem caused by conventional Sliding mode control, a fuzzy sliding mode controller with adaptive switching gain adjustment function is designed. Finally, a joint simulation platform for unmanned vehicle path tracking was built based on Simulink/Carsim software.

In order to improve the tracking reliability of unmanned vehicles, a real-time tracking method for unmanned vehicle paths based on electronic tags is proposed.

2 Path Information Collection of Unmanned Logistics Vehicles Based on Electronic Tags

Warehouse identification number of a supply chain member. A supply chain member can have multiple warehouses, and the warehouse identification number is used to uniquely identify a warehouse [6–8]. Warehouse is the place where products are stored, and also the place where readers identify product information. Usually, a reader is placed at the entrance and exit of the warehouse respectively to capture the warehousing and discharging time of products, as shown in Fig. 1.

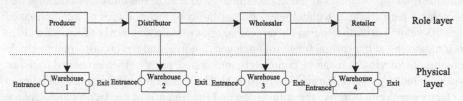

Fig. 1. Path node of unmanned logistics vehicle

The basic principle of product logistics path tracking based on electronic labels is to set up readers and writers at the entrances and exits of warehouses of supply chain members. When the product manufacturer produces the product, they attach a unique product electronic code to the product and write it into the attached electronic label. When the product with the electronic label passes through the entrances and exits of these warehouses, the reader and writer will automatically capture the product identification code, And record the system time when the product was identified, thus forming an information node that records the product logistics path. A series of information nodes arranged in chronological order form the entire logistics path that the product has passed through. Decision makers can use this path information to query and track the historical status of the product, and also know the current location and status of the product [9–11]. The structure of a typical product logistics path tracking system based on electronic tags is shown in Fig. 2.

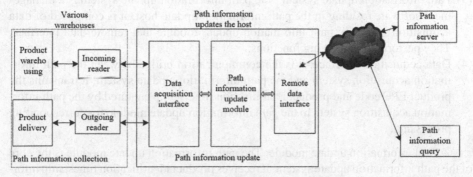

Fig. 2. Principle of path tracking data acquisition based on electronic tag

It consists of four functional modules: path information collection system, path information update system, path information storage system and path information query system. Its functions are as follows:

(1) Path information collection system. The core of the path information collection system is product coding and product identification. In logistics path tracking systems based on electronic tags, EPC codes are usually used as the unique identification codes for products. The EPC code Auto-ID research center proposes a coding standard for electronic labels, which enables all goods worldwide to have a unique identifier, and its biggest feature is the ability to identify individual products. The product identification system includes electronic labels and readers. Each product is accompanied by an electronic label, which contains the EPC code as the unique code of the product. When electronic tags with EPC codes are stored in the sensing area of the reader, the EPC code will be automatically captured by the reader, achieving automated product identification. The system time when the product is identified is combined with the reader number, reader position, warehouse number, and supply chain member angle to form a logistics path information node. According to different purposes, readers can be divided into inbound readers and outbound readers, and the corresponding logistics path information nodes generated are entry information nodes and exit information nodes, respectively [12–14]. The functions of inbound and outbound readers are as follows:

① Incoming reader: it is set at the entrance of the warehouse to automatically identify the products entering the warehouse, automatically record the incoming time of the products, provide information collection support for generating the entry information node, and transmit to the path information update module through the data collection interface for corresponding processing;

(2) Outgoing reader: set at the warehouse exit, automatically identifies the products leaving the warehouse, automatically records the time of the products leaving the warehouse, provides information collection support for generating the export information node, and transmits to the path information update module through the data collection interface for corresponding processing [15].

(3) Path information update system. The path information update system is a management software residing in the path information update host. It is composed of data acquisition interface, path information update module and remote data interface. They perform the following functions:

(4) Data acquisition interface: it is the communication bridge between the path information acquisition system and the path information update system. It transmits the product EPC code and product identification time stamp captured by the path information acquisition system to the path information update module for corresponding processing;

② Path information update module: The path information update module is the core of the path information update system. It receives product identification timestamp information from the path information collection system through a data collection interface, generates entry information nodes or exit information nodes based on the purpose of the reader and writer, and then updates the path information record of the path information storage system through a remote data interface to expand new information nodes. When the product is recognized by different readers and writers, new path information nodes continue to expand and path information records constantly refresh. The process of updating path information is shown in Fig. 3.

③ Remote data interface: It is the bridge of communication between the path information update system and the path information storage system. Because the path information is stored in an independent product path information server in a standard XML format, the path information update system and the path information storage system exchange information through the client/server mode. The path information server usually provides standard remote access interfaces, such as SOAP interface, which allows message exchange between different applications through HTTP communication protocol and in XML format. Therefore, the path information update system of various platforms can update path information through standard remote data interface, which has good platform independence.

(3) Path information storage system. The path information is stored in a standard XML format on an independent path information server (usually also known as a PML server), which is generally established and maintained by the product manufacturer and authorized to downstream supply chain members for path updates. Usually, a good path information storage system should have two conditions: ① standardization of path information storage format; ② Standardization of path information access interfaces. This is achieved through XML and SOAP interfaces, which complement each other and together provide a standard set of information access patterns for clients.

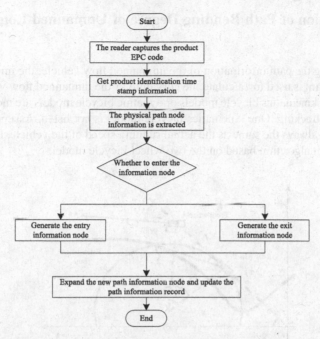

Fig. 3. Path information update process

(4) Path information query system. The path information query system is a functional part that directly provides product path information support and decision-making basis for supply chain members or decision-makers. For a product logistics path tracking system, its supported query functions are as follows:

① Full path query: query the information of all path nodes that the product passes through from production to the current state;

(2) Path query in a certain time period: query the information of the path nodes passed by the product in a specified time interval;

(3) The upstream member query of the supply chain: the information of the upstream member node of a certain member node in the supply chain is logically realized by determining all the leading nodes of the node.

(4) The downstream member query of the supply chain: the query of the downstream member node information of a certain member node in the supply chain is realized logically by determining all subsequent nodes of the node;

⑤ Product direct source query: the query of the product direct source node of a member node in the supply chain is realized logically by determining the close preceding node of the node;

⑥ Direct flow of products query: The query of the direct flow of products of a member node in the supply chain is logically realized by determining the close node of the node.

3 Calculation of Path Bending Degree of Unmanned Logistics Vehicle

After collecting the path information of the unmanned flow vehicle, the improved Pure Pursuit algorithm is used to calculate the curvature of the unmanned flow vehicle path.

At present, kinematics bicycle models or dynamic bicycle models are mostly used in real-time path tracking. One kinematics bicycle model is not listed. Assuming that the wheel speed is always the same as the actual running speed of the vehicle, Fig. 4 shows the Pure Pursuit algorithm based on the two wheel bicycle model.

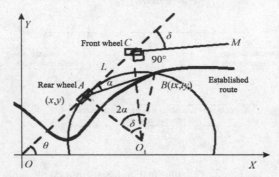

Fig. 4. Pure Pursuit algorithm based on the two-wheeled bicycle model

As shown in Fig. 4, the vehicle is assumed to be A two-wheel bicycle model, and the front and rear wheels are represented by C and A respectively. The model approximately describes its motion at low speed and medium Angle, and can be used to update its motion state. Among them, the update equation of position (x, y) and heading Angle (θ) of the unmanned logistics vehicle is:

$$\begin{bmatrix} \dot{x} \\ \dot{y} \\ \dot{\theta} \end{bmatrix} = \begin{bmatrix} v \cos \theta \\ v \sin \theta \\ \frac{v \tan \delta}{L} \end{bmatrix} \tag{1}$$

In the formula, L represents the wheelbase between the front and rear wheels AC of the unmanned flow vehicle, determined based on actual measurement values. v represents the current speed of the unmanned flow vehicle, and δ represents the angle of rotation in the current heading direction of the unmanned flow vehicle. Counterclockwise is positive, and clockwise is negative.

In triangle AOC, using the trigonometric function relationship, it can be known that the rotation angle is:

$$\delta = \arctan \frac{L}{R} \tag{2}$$

In the formula, R represents the radius of the unmanned logistics vehicle when turning.

In the Pure Pursuit algorithm, the rear-wheel steering motion of unmanned logistics vehicles is approximated as a part of a circle. In triangle AOC, sine theorem is used to show that:

$$\frac{L_f}{\sin(2\alpha)} = \frac{R}{\sin(2\pi - \alpha)} \tag{3}$$

In the formula, L_f represents the forward looking distance between the current position A of the unmanned vehicle and the target point B, while α represents the angle at which the unmanned vehicle turns from the current heading direction (AC) towards the target point direction (AB), which is the heading angle deviation. Counterclockwise is positive, and clockwise is negative.

From the above formula, it can be seen that the turning angle of a pedestrian free vehicle is:

$$\delta = \tan^{-1}\left(\frac{2L\sin\alpha}{L_f}\right) \tag{4}$$

For local areas, the specific value of α needs to be analyzed according to the direction between the current position of the unmanned logistics vehicle and the target point of the predetermined path, and the specific relationship is shown in Fig. 5.

As shown in Fig. 5, if the unmanned vehicle position at a certain moment is (x, y) and the target point coordinate is (tx, ty), then β can be represented as:

$$\beta = \tan^{-1}\frac{ty - y}{tx - x} \tag{5}$$

In the formula, β represents the Angle between the line between the current position A and the target point B and the positive direction of axis x.

According to Fig. 5 (a) and 5 (b), when the target point is located on the left side of the trajectory, the relationship between α, β and θ is:

$$\begin{cases} \alpha = \beta - \theta - \pi, \theta \in (-\pi, \beta] \\ \alpha = \beta - \theta + \pi, \theta \in [\beta, \pi] \end{cases} \tag{6}$$

As shown in Fig. 5 (c), when the target point is located at the upper right of the trajectory, the relationship between α, β, and θ is:

$$\begin{cases} \alpha = \beta - \theta, \theta \in (\beta - \pi, \pi] \\ \alpha = \beta - \theta - 2\pi, \theta \in [-\pi, \beta - \pi] \end{cases} \tag{7}$$

According to Fig. 5 (d), when the target point is located at the lower right of the trajectory, the relationship between α, β and θ is:

$$\begin{cases} \alpha = \beta - \theta, \theta \in (-\pi, \pi + \beta] \\ \alpha = \beta - \theta + 2\pi, \theta \in [\pi + \beta, \pi] \end{cases} \tag{8}$$

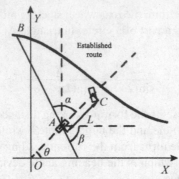

(a) The target point is located at the top left of the trajectory

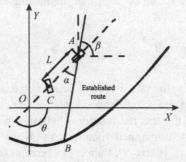

(b) The target point is located at the bottom left of the trajectory

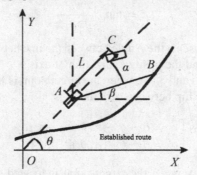

(c) The target point is located at the top right of the trajectory

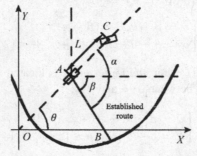

(d) The target point is located at the bottom right of the trajectory

Fig. 5. Interrelationship between target points and trajectories

Through the above calculation, the heading angle deviation α of the unmanned flow vehicle can be obtained. According to formula (4), the steering wheel angle δ can be calculated, and the steering wheel angle command can be transmitted to the executing mechanism to control the angle corresponding to the steering wheel angle.

In order to ensure the smoothness and continuity of speed changes of unmanned flow vehicles on paths with significant curvature changes, an improved algorithm for forward looking distance is proposed based on existing theories, which can be expressed as:

$$\begin{cases} L_f = \frac{1}{2a_{max}}v^2 + k_3 v + L_{fc} \\ v = \frac{v_{max} - v_{min}}{(C_2 - C_1)^2}(C_2 - C_1)^2 + v_{min} \end{cases} \tag{9}$$

In the formula, a_{max} is the maximum braking acceleration, k_3 is the speed coefficient, L_{fc} is the initial looking distance, where the minimum turning radius of the unmanned logistics vehicle is taken, C_1 is the maximum bending degree of a given path, C_2 is the minimum bending degree of a given path, v_{min} is the minimum speed of the unmanned logistics vehicle set in advance, and v_{max} is the maximum speed of the unmanned logistics vehicle set in advance.

Formula (9) can reduce the speed and forward looking distance as the road curvature increases, and achieve continuous changes in speed without the need to set the speed into high, medium, and low gears based on experience for the curvature range. The improved algorithm can achieve adaptive changes in speed with the curvature of a given path, based on the maximum and minimum values of the driving speed of a pedestrian free vehicle.

The method for determining the target point is shown in Fig. 6.

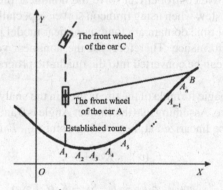

Fig. 6. Target point determination method

Given A predetermined path, firstly, the nearest point from the predetermined path is selected as A_1 based on the current position of the unmanned logistics vehicle. Then, the adjacent points behind point A_1, which is consistent with the forward direction of the unmanned logistics vehicle, are added successively. When the desired sum is greater than the corresponding forward distance L_f at the current speed, the point is selected as the target point B.

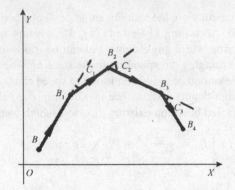

Fig. 7. Determination method of bending degree

The determination method of bending degree is shown in Fig. 7.
The solution formula for curvature C can be expressed as:

$$C = \sum_{i=1}^{3} c_i \tag{10}$$

In the formula, c_i is the Angle between two adjacent broken lines.

4   Real Time Tracking of Traffic Paths Without People

Nonlinear model predictive controller can solve the nonlinear problems of the system, but the solution speed is slow when using fmincon solver, especially when the prediction time domain and control time domain are long, nonlinear model predictive control will lose large real-time performance. Therefore, for the nonlinear vehicle model, the linearization optimization can be converted into the quadratic programming problem with faster solving speed.

In this section, the magic formula of the tire is used in the analysis of the dynamics of the unmanned vehicle tire. Assuming that the camber Angle is 0 and the tire is unchanged, the magic formula can be linearized at the working point $(\alpha_0, F_0(\alpha))$:

$$F_c(\alpha) = K\alpha + \lambda \tag{11}$$

In the formula, $K = \frac{dF_c}{d\alpha}$ and $\lambda = F_0 + K \cdot S_h - K(\alpha + S_h)$.

Considering the constant speed of unmanned vehicles and considering the assumption of small front wheel angles, a 2 degree of freedom model for lateral and yaw is established using Newton's second law:

$$\begin{cases} m(\ddot{y} + v_x\dot{\varphi}) = 2F_{ef} + 2F_{cr} \\ I_z\ddot{\varphi} = 2aF_{cr} - 2bF_{cr} \end{cases} \tag{12}$$

In the formula, a and b are the distance from the center of mass of the unmanned logistics vehicle to the front and rear axle, m is the mass of the vehicle, I_z is the moment of inertia of the unmanned logistics vehicle around axis z, F_{ef} and F_{cr} are the lateral force of the front and rear wheels, y is the lateral displacement of the center of mass of the unmanned logistics vehicle, v_x is the longitudinal velocity of the center of mass of the unmanned logistics vehicle, and φ is the yaw Angle.

By substituting formula (11) into formula (12), we can get:

$$\begin{cases} \ddot{y} = -v_x\dot{\varphi} + \frac{2}{m}\left[K_{cf}\left(\delta_f - \frac{\delta_f}{v_x}\right) - K_{cr}\frac{\dot{y}-b\dot{\varphi}}{v_x}\right] + \frac{2\lambda_{cf}+2\lambda_{cr}}{m} \\ \ddot{\varphi} = \frac{2a}{I_z}\left[K_{cf}\left(\delta_f - \frac{\dot{y}+a\dot{\varphi}}{v_x}\right)\right] + \frac{2b}{I_z}K_{cr}\frac{\dot{y}-b\dot{\varphi}}{v_x} + \frac{2(\lambda_{cf}a-\lambda_{cr}b)}{v_x} \end{cases} \tag{13}$$

In the formula, K_{cf}, K_{cr}, λ_{cf}, and λ_{cr} represent the parameters K and λ after linearizing the front and rear wheel lateral forces.

The Taylor series is used to expand at the working point G, ignoring the higher-order term, and the following results are obtained:

$$\dot{Y} = \dot{y}\cos\varphi_r + v_x\sin\varphi_r + v_x\varphi\cos\varphi_r - \dot{y}_r\varphi\sin\varphi_r \tag{14}$$

Thus, the whole unmanned logistics vehicle system can be written as an incremental linear time-varying state space expression:

$$\begin{cases} \dot{x}(t) = A(t) + B(t)u(t) + D(t) \\ y(t) = Cx(t) \end{cases} \tag{15}$$

For this continuous state space equation, the first order difference quotient method is used for discretization to obtain the discrete state space expression:

$$\begin{cases} x(k+1) = A(k)x(k) + B(k)u(k) + D(k) \\ y(k+1) = Cx(k+1) \end{cases} \tag{16}$$

In the formula, $A(k) = I + TA(t)$, $B(k) = TB(t)$, $D(k) = TD(t)$ and T represent sampling time.

Suppose:

$$\xi(k) = \begin{bmatrix} x(k) \\ u(k-1) \end{bmatrix} \tag{17}$$

Change formula (16) to incremental model:

$$\begin{cases} \dot{\xi}(k+1) = \tilde{A}(k)\xi(k)\Delta u(k) + \tilde{D} \\ \eta(k+1) = \tilde{C}(k)\xi(k) \end{cases} \tag{18}$$

According to the basic idea of model predictive control, by setting control time domain N_c and prediction time domain N_p, and $N_c \leq N_p$, the output in the prediction time domain can be written as:

$$Y(k+1) = \begin{bmatrix} \eta(k+1) \\ \eta(k+2) \\ \vdots \\ \eta(k+N_p) \end{bmatrix} \tag{19}$$

Write the input quantity in time domain N_c as a vector:

$$\Delta U(k) = \begin{bmatrix} \Delta u(k) \\ \Delta u(k+1) \\ \vdots \\ \Delta u(k+N_c-1) \end{bmatrix} \tag{20}$$

According to the model predictive control principle, the system output equation at time k can be obtained:

$$Y(k+1) = S_t \xi(k) + S_u \Delta U(k) + \Phi \tilde{D} \tag{21}$$

Due to the complexity of the unmanned vehicle dynamics model itself, a relaxation factor was added to the design of the objective function, which is:

$$J(\xi(t), u(t-1), \Delta U(t)) = \sum_{i=1}^{N_p} \|\Delta \eta(t+i)\|_Q^2 + \sum_{i=1}^{N_c-1} \|\Delta u(t+i)\|_R^2 + \rho \varepsilon^2 \tag{22}$$

In the formula, $\Delta \eta(t+i)$ is the deviation between the predicted output and the reference trajectory in the forecast time domain; ε is the relaxation factor; ρ is the weight coefficient of the relaxation factor; Q is the weight coefficient of the output deviation; R is the weight coefficient of the control increment.

Suppose:

$$E(t) = S_\xi \xi(k) + \Phi \tilde{D} - Y_{ref} \tag{23}$$

By substituting formula (22), it can be converted into a standard Quadratic programming problem with fast solution speed:

$$J(\xi(t), u(t-1), \Delta U(t)) = \left[\Delta U(t)^T \varepsilon\right]^T H(t) \left[\Delta U(t)^T \varepsilon\right] + G(t) \left[\Delta U(t)^T \varepsilon\right] + p(t) \tag{24}$$

Each step of model predictive control can also be transformed into a quadratic programming problem with constraints:

$$\min J(\xi(t), u(t-1), \Delta U(t))$$

$$s.t. \begin{cases} \Delta U_{min} \leq \Delta U_t \leq \Delta U_{max} \\ U_{min} \leq A \Delta U_t + U_t \leq U_{max} \\ Y_{h,min} \leq Y_h \leq Y_{h,max} \\ Y_{s,min} - \varepsilon \leq Y_s \leq Y_{s,max} + \varepsilon \end{cases} \tag{25}$$

In each solution cycle, a series of optimal control output increment sequences in the control time domain can be obtained by calculating the above Quadratic programming problem:

$$\Delta U_t^*(t) = \left(\Delta u_t^*, \ldots, \Delta u_{t+N_c-1}^*\right) \tag{26}$$

According to the principle of model predictive control, only the first element of the control sequence is taken as the actual input increment of the controlled object, then the actual input of the controlled object at the next moment is:

$$u(t) = \Delta u_t^* + u(t - 1) \tag{27}$$

After entering the next cycle, repeat the above process to make the unmanned vehicle track the target path.

5   Experimental Verification

The experiment platform of unmanned logistics vehicle was constructed, with the maximum braking acceleration of 3 m/s^2 and the speed coefficient of 0.2 m. The minimum turning radius is 5 m. Carry out data collection along an annular path, and then obtain the change of the established path curvature of the unmanned logistics vehicle. Then, according to the position relationship between the unmanned logistics vehicle and the predetermined path, the proposed method and the reference method [4] Udine are respectively adopted to track the established path, and the starting position of the unmanned logistics vehicle is set as (523350 m, 4057250 m), the initial heading Angle is 0°, the wheelbase of the unmanned logistics vehicle is 1 m, the sampling time is 1 s, the maximum traffic jam is 30 km/h, and the minimum speed is 5 km/h.

The equipment required for real-time tracking of the flow car trajectory without characters is shown in Table 1.

Table 1. Equipment used

Device	Model
GPS	Quectel L80
Vehicle sensor	Gyroscope for STMicroelectronics
Communication devices	Sierra Wireless AirPrime EM7455
Data storage device	SSD hard disk
Tracking system	Google Maps

In order to analyze the performance of the improved Pure Pursuit algorithm, the response speed of the algorithm was used as an indicator to verify the performance of the Pure Pursuit algorithm before and after the improvement.

From the comparison results of response speed shown in Fig. 8, it can be seen that the improved Pure Pursuit algorithm has significantly improved response speed, with a maximum response time of no more than 1.0 s. Therefore, it indicates that the improvements made in this article improve the response speed of the Pure Pursuit algorithm.

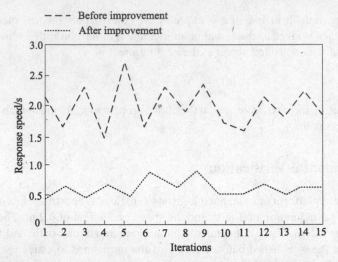

Fig. 8. Response speed of Pure Pursuit algorithm before and after improvement

The path tracking accuracy and lateral tracking error results of the method in this paper and the method in reference [4] are shown in Fig. 9 and 10 respectively.

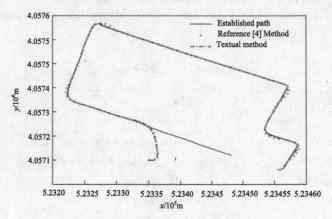

Fig. 9. Path tracking accuracy

From the results shown in Figs. 9 and 10, it can be seen that compared to the method in reference [4], the path tracked by this method is closer to the established path, and the lateral tracking error is smaller. Therefore, it is demonstrated that the method proposed in this paper can improve the real-time path tracking accuracy of unmanned vehicles.

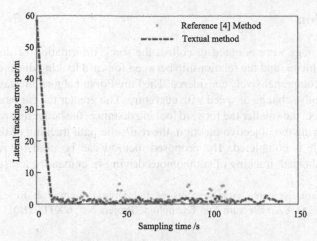

Fig. 10. Lateral tracking deviation

In order to further accurately analyze the tracking performance of the method proposed in this paper, the tracking time of the method proposed in this paper was compared with that of the method proposed in reference [4]. The comparison results of tracking time are shown in Table 2.

Table 2. Tracking time results

Number of experiments	Tracking time consumption/s	
	Proposed method	Reference [4] method
1	1.36	8.96
2	1.05	7.68
3	1.14	9.16
4	1.34	10.56
5	1.21	8.67

From the tracking time results shown in Table 2, it can be seen that under multiple comparative experiments, the tracking time of our method is significantly lower than that of the reference [4] method. The longest detection time of our method is 1.36 s, while the longest tracking time of the reference [4] method is 10.56 s. Therefore, it is demonstrated that the method proposed in this paper can achieve fast tracking of unmanned traffic paths.

6 Conclusion

The electronic tags were adopted to collect the track information of unmanned and vehicle-free vehicles, and the relationship between forward looking distance, speed and curvature was comprehensively considered. The Pure Pursuit algorithm was improved to realize the adaptive change of speed with curvature. The greater the curvature, the lower the vehicle speed, the smaller the forward looking distance, the better the tracking effect. Finally, by solving the objective function, the real-time path tracking of the unmanned logistics vehicle is completed. The proposed method can be used for reference and application in the path tracking of autonomous driving in unmanned logistics.

Acknowledgement. Research Achievements of the Digital Innovation Research Team of Chongqing College Of Architecture And Technology (Project No.: CXTD22B01).

References

1. Prokopyev, I.V., Sofronova, E.A.: Study on control methods based on identification of unmanned vehicle model. Procedia Comput. Sci. **186**(15), 21–29 (2021)
2. Heo, J., Ha, J., Lee, J., et al.: Dynamic window approach with path-following for unmanned surface vehicle based on reinforcement learning. Korea Inst. Mil. Sci. Technol. **24**(1), 146–151 (2021)
3. Yoshimoto, Y., Watanabe, T., Nakamura, R., et al.: Effectiveness of digital twin computing on path tracking control of unmanned vehicle by cloud server. IEICE Trans. Commun. **16**(8), 105–109 (2022)
4. Wang, X., Ling, M., Rao, Q., et al.: Research on fusion algorithm of unmanned vehicle path tracking based on improved Stanley algorithm. Automobile Technol. **07**, 25–31 (2022)
5. Li, Z., Wang, Y., Yin, K., et al.: Design of path following controller for unmanned vehicle based on FSMC. J. Hebei Univ. Water Resour. Electr. Eng. **32**(01), 9–15 (2022)
6. Wang, Z., Zhang, H., Zhang, H., et al.: Adaptive event based predictive lateral following control for unmanned ground vehicle system. Int. J. Robust Nonlinear Control **31**(10), 4744–4763 (2021)
7. Mao, X., Yu, B., Shi, Y., et al.: Real-time online trajectory planning and guidance for terminal area energy management of unmanned aerial vehicle. J. Ind. Manag. Optim. **19**(3), 1945–1962 (2023)
8. Li, J., Sun, T., Huang, X., et al.: A memetic path planning algorithm for unmanned air/ground vehicle cooperative detection systems. IEEE Trans. Autom. Sci. Eng. **32**(9), 1–14 (2021)
9. Strawa, N., Ignatyev, D.I., Zolotas, A.C., et al.: On-line learning and updating unmanned tracked vehicle dynamics. Electronics **10**(2), 187–193 (2021)
10. He, Z., Dong, L., Sun, C., et al.: Asynchronous multithreading reinforcement-learning-based path planning and tracking for unmanned underwater vehicle. IEEE Trans. Syst. Man Cybern. Syst. **41**(19), 1–13 (2021)
11. Wang, Z., Xin, P., Sun, H., et al.: Unmanned vehicle path tracking based on contraction constraint model predictive control. Control Decis. **37**(03), 625–634 (2022)
12. Kou, F., Yang, H., Zhang, X., et al.: A lateral control strategy for unmanned vehicle path tracking using state feedback. Mech. Sci. Technol. Aerosp. Eng. **41**(01), 143–150 (2022)
13. Li, Y., Guo, C., Li, Y., et al.: Trajectory tracking control of autonomous vehicle based on steering and braking coordination. Syst. Eng. Electron. **45**(04), 1185–1192 (2023)

14. Qiu, P., Wang, G., Zhao, L., et al.: Unmanned vehicle path planning based on structured road improved artificial potential field method. Mach. Des. Manuf. **03**, 291–296 (2023)
15. Wu, Y., Liang, H.: Research on obstacles avoidance control method of unmanned ground vehicles based on adaptive model prediction. Manuf. Autom. **45**(02), 198–202 (2023)

Error Motion Tracking Method for Athletes Based on Multi Eye Machine Vision

Yanlan Huang[1]([✉]) and Chunshou Su[2]

[1] Guangxi College for Preschool Education, Nanning 530022, China
hy15825@163.com
[2] Yulin Normal University, Yulin 537000, China

Abstract. Traditional methods are unable to perform three-dimensional detection of erroneous movements, resulting in insufficient accuracy in tracking athlete erroneous movements. Therefore, the tracking method for athlete erroneous movements based on multi eye machine vision is highlighted. Using multi eye machine vision technology to construct an athlete error motion tracking framework and obtain athlete error motion image timing. From the perspective of regional consistency and similarity, segment machine vision images of athlete's incorrect actions. Apply Canny operator to detect athlete's incorrect actions, obtain pixel values of edge images, and remove false edges. The design is based on a multi eye machine vision athlete error action recognition process, obtaining unknown vectors. With the support of a multi eye machine vision detection system, the absolute value of brightness difference between two frames of images is calculated, and the Hom Schunck algorithm is combined to track the optical flow field to achieve athlete error action tracking. From the experimental verification results, it can be seen that the tracking curve of this method for three types of erroneous actions is consistent with the actual curve, and the maximum tracking accuracy is 93%, which can accurately track athlete's erroneous actions.

Keywords: Multi Eye Machine Vision · Athlete Error Movement Tracking · Hom-Schunck Algorithm · Error Action Image

1 Introduction

A high-quality sports event is often inseparable from the tactical arrangement of the coach. Whether the real-time match information of athletes, such as speed and exact route, can be well obtained plays a decisive role in the coach's arrangement of personnel and tactics. Real time access to athletes' competition information is the core function of athletes' tracking system. With the continuous improvement of sports technology, in the process of sports technology research, the demonstration of correct movement has become a key topic in the teaching field. In the actual teaching process, due to the obvious difference between the students' cognitive and understanding level and their action ability, some students have more wrong actions, and they are slow to master correct sports actions. In this state, how to effectively correct wrong movements in sports has

L. Yun et al. (Eds.): ADHIP 2023, LNICST 550, pp. 150–164, 2024.
https://doi.org/10.1007/978-3-031-50552-2_10

become the main problem to be solved in this field. In recent years, with the promotion of computer image processing technology in China, computer vision feature analysis and image processing technology have been widely used in the analysis of human body structure, which can analyze various forms of human body in motion. In this context, the field of sports has also begun to introduce computer vision feature analysis technology to athletes' action recognition and correction, so as to improve the effectiveness and judgment of athletes' training.

In outdoor sports events, wearable GPS devices are often used to achieve real-time tracking of athletes. In indoor sports events, high-precision multi-sensor systems are generally used to obtain the location information of athletes. The above two methods will make the tracking system too complex and difficult to maintain. In view of the short-comings of existing tracking methods, a tracking method that can be applied to mobile devices is proposed. Reference [1] proposed a tracking method based on the stacked short-term memory network. This method takes the original track of athletes' actions as input, uses the stacked short-term memory network to learn the feature representation of space-time window, and cascades the implicit spatial representation of athletes on the field through an additional full connection layer. The Softmax layer is used to estimate the probability of athletes' final actions. Each final action is associated with an expected score, and it is used to estimate the expected score to obtain the tracking results of athletes' wrong actions; Reference [2] proposed a tracking method based on LSTM neural network, which smoothed and denoised the original 3D skeleton data on the basis of 3D skeleton data to conform to the smooth rule of human joint movement. A fusion feature composed of static features and dynamic features is constructed to represent human actions, and a key frame extraction model is introduced to extract key frames in human action sequences to reduce the amount of computation. The human motion classification model of Bi-LSTM neural network based on LSTM neural network is established. Attention mechanism and Dropout are introduced to classify and recognize human motion. However, the two methods mentioned above can be well expanded and have strong applicability, which can meet the needs of somatosensory interactive applications, but the recognition accuracy is very low. Therefore, an error motion tracking method for athletes based on multi eye machine vision is proposed.

2   Tracking Image Processing Based on Multi Eye Machine Vision

The multi eye machine vision technology is used to build the athlete error motion tracking framework, and its structure is shown in Fig. 1.

It can be seen from Fig. 1 that the tracking target is determined according to the framework, and the end to end training is conducted through forward and reverse propagation to obtain the final tracking result [3]. Convolution operation is a sparse, parameter sharing, variable operation. In the convolution operation, a filter is used to scan the two-dimensional image as a whole (filter). Each pixel in the vector graph will be scanned on the same convolution kernel, so its convolution kernel is the weight. Weight assignment makes the deep convolution network only need to learn a group of parameters, thus greatly reducing the number of parameters.

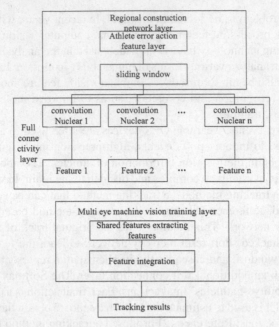

Fig. 1. Tracking framework of multi eye machine vision

2.1 Time Sequence Acquisition of Athletes' Wrong Action Images

Obtaining athletes' action images is the premise of extracting wrong action features. Due to the large visual error of action images obtained by traditional methods, it directly affects the accuracy of extracting wrong action features [4]. For this reason, the multi eye machine vision system is used to obtain athletes' action images. The components used to obtain images with the OV7670 camera are PL and PS, and the athletes' action images are displayed using the VGA interface according to the data integrity of line interruption and field interruption. The VGA timing is shown in Fig. 2.

In the multi eye machine vision system, the OV7670 camera is selected to obtain the movement image of athletes using the camera [5]. The OV7670 camera is a component of the COMOS camera and has the ability to acquire color images. The photosensitive array can achieve a maximum transmission rate of 640 * 680 and a maximum transmission rate of 30 frames/second. The camera has only one set of parallel data interfaces, marked as Y [7:0]. The pixel values of action images are read through the data interface, and the action images of football players are obtained in parallel.

2.2 Machine Vision Image Segmentation of Athletes' Wrong Actions

From the perspective of regional consistency and similarity, it can effectively prevent the integrity of region extraction caused by changes in image grayscale; Secondly, from the perspective of threshold segmentation, the time required for this method can be shortened [6]. Use the k-means clustering algorithm to converge the sample points to the position with the highest probability density, thereby obtaining regions with similar grayscale.

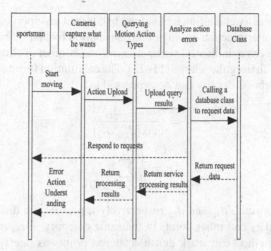

Fig. 2. Time sequence of athletes' action images

When applied to motion error machine vision image segmentation, the k-means clustering algorithm has advantages and characteristics such as simple, efficient, intuitive clustering results, applicability to large-scale datasets, and strong interpretability. This algorithm can quickly calculate the distance between sample points and the cluster center, allocate sample points to the nearest cluster center, and iteratively update the position of the cluster center, thereby segmenting erroneous action images into different clusters and intuitively distinguishing different types of errors.

The probability density value of similar regions is calculated using the formula:

$$\rho(a) = \frac{1}{kD} \sum_{x_l \in s_D} f\left(\frac{a}{D}\right) \tag{1}$$

In formula (1), a represents the sampling point; s_D represents the area with a fixed bandwidth of D; $f(\cdot)$ represents a symmetric kernel function; k represents the number of calculations [7–9]. After determining the initial point, use, k-means clustering algorithm to perform iterative processing according to the following steps:

Step 1: Use the density clustering method to classify the athletes' error movement tracking data, and divide the normal data and error data.In the athlete's error action tracking data set, error data contains some feature quantities that must be preprocessed [10]. The preprocessed training data set can be regarded as a feature matrix, in which the data in the same row is consistent with the features of multi eye machine vision. For the clustering problem in multidimensional vector space, the density of attributes is used to cluster each attribute, and the form of adjacent regions of eigenvalues can be expressed as:

$$q(z) = \{z_1, z_2 \in V_m | \ dist(z_1, z_2) \le \vartheta\} \tag{2}$$

In formula (2), z_1 and z_2 represent two eigenvalue objects; V_m represents a set of m eigenvalues; ϑ represents the core object in the cluster.

Step 2: Using density clustering technology to conduct unsupervised machine learning method, it can determine the density of its distribution by taking the size and concentration of adjacent areas as indicators without setting the number of clusters in advance, so as to find clusters with irregular shapes [11–13]. The calculation formula of neighborhood radius and neighborhood density can be expressed as:

$$l = \frac{d_{z_2} - d_{z_1}}{\max(d_{z_2}, d_{z_1})} \tag{3}$$

$$\rho = \frac{1}{n} \sum_{k=1}^{n} l_k \tag{4}$$

In the above formula, d_{z_1} and d_{z_2} respectively represent the distance between the two samples z_1 and z_2 and other points in the same category; n represents the number of samples. This method can mark dense scattered points as one type, and scattered scattered points as another type to distinguish between normal data and error data [14, 15].

Step 3: Continuously iterate and set the convergence threshold λ. The constraint conditions for convergence values can be expressed as:

$$\|a_{n+1} - a_n\| < \lambda \tag{5}$$

When the calculation result of formula (5) is satisfied, it means that the convergence density reaches the maximum value, otherwise, it cannot. This clustering method can avoid the problem of region segmentation errors.

2.3 Tracking Image False Edge Processing

The edge of an image is the grayscale value space boundary in which the pixel neighborhood has a stepping property. Extracting a set of pixels with significant grayscale changes from the image is the result of image edge detection. The Canny operator is a first-order derivative Gaussian method that accurately extracts image edges from an image. By analyzing image gradients, the edges of each pixel in the image can be obtained. The main reasons why Canny operator can accurately detect edges when extracting image edges are as follows. Firstly, the Canny operator adopts multi-stage operations, including Gaussian filtering, gradient calculation, and non maximum suppression, which can effectively suppress noise, locate edges, and refine edges. Secondly, the Canny operator uses Gaussian filtering to smooth the image, reducing noise interference and making subsequent edge detection more stable and reliable. Then, by calculating the gradient size and direction of each pixel, the Canny operator can locate the edge regions in the image, as the pixel values at the edges change greatly, and the gradient values also increase accordingly. Next, the Canny operator uses non maximum suppression to refine edges, preserving local maxima in the gradient direction, and further improving the accuracy of edge detection. Finally, the Canny operator uses dual threshold processing to filter edges, identifying pixels with gradient values greater than the high threshold as strong edges, while pixels between the low and high thresholds are identified as weak edges.

By connecting strong edges with adjacent weak edges, the Canny operator can obtain the final accurate edge result. Therefore, the Canny operator can extract accurate edge information from images through multiple stages of processing and parameter settings. Canny operator is applied to detect athletes' wrong actions, and the pixel value of edge image obtained can be expressed as:

$$\beta'(a, b) = \Delta\beta(a, b) \otimes \Delta c(a, b) \tag{6}$$

In formula (6), $\Delta\beta(a, b)$ represents the loss compensation pixel value; $\Delta c(a, b)$ represents the pulse response compensation value. In order to solve the problem of false edges, a weighted steering filter is constructed by combining the steering filter and edge sensing weight, which can strengthen the low-frequency components in the image and prevent false edges from appearing in the image. The calculation formula of edge perception weight is:

$$\omega(i) = \frac{1}{k} \sum_{i=1}^{k} \frac{S^2(i)}{S^2(I)} \tag{7}$$

In formula (7), S represents the variance of pixel point i as the guide image I within the pixel neighborhood. The pixel value of the image smoothing position is different from that of the edge position, where the edge perception weight of the smoothing position is less than 1 and the edge position is greater than 1. In the image, the two nearest edges are taken as true edges, which are very different from the average value of each pixel. For other pixels, if the pixel value jumps too large, it is called false edge. The false edge is determined according to the dispersion of each pixel in the image. The smallest of the two pixels is selected as the true edge, and the remaining pixels are false edges.

3 Identification and Tracking of Athletes' Wrong Movements

3.1 Athlete Error Action Recognition Based on Multi Eye Machine Vision

The iterative algorithm can quickly calculate the local maximum probability of the target to adapt to the multi-target deformation. Once the target is occluded, the number of multi-target maximum points will become more. At this time, the segmented multi-target tracking and positioning results will be biased, and the coordinate information will be lost. Segmented multi-target tracking and location method can track the target well for the problem of target occlusion, but the selection of location information is relatively strict.

Compared to other factors, multi eye machine vision systems need to focus on occlusion issues when applied. Occlusion issues can lead to information loss, target deformation, and target loss. Therefore, when using multi eye machine vision systems for multi object tracking, the influence of other factors should be ignored and occlusion issues should be emphasized.

In the whole tracking process, the feature information needs to be fused according to Bhattacharyya coefficient, and the target model is established for the current frame through Dempster-Shafer evidential reasoning. After iterative processing, candidate targets are captured, and similarity judgment is made according to the selected targets. If the similarity is greater than the set threshold, it means that the multiple targets are not blocked, and the multi target tracking is continued using the multi eye machine vision system; On the contrary, if the similarity is less than the set threshold, it means that multiple targets are occluded. At this time, the target template should not be updated, and the Dempster-Shafer evidential reasoning method should be directly switched to track. The method of comparing similarity with threshold can ensure good tracking effect when multiple targets are not occluded. Once the target is occluded, the method can enhance the effectiveness of tracking.

The process of athlete error action recognition based on multi eye machine vision is shown in Fig. 3.

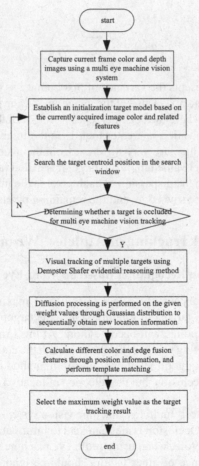

Fig. 3. Athlete's Error Action Recognition Process Based on Multi eye Machine Vision

It can be seen from Fig. 3 that the camera orientation is calculated by the coordinate information obtained from the multi object tracking and positioning based on multi eye machine vision and the corresponding points as well as the internal parameters of the multi eye machine vision system, the tracking target position is estimated by Dempster-Shafer evidential reasoning method, and the virtual control points are used to represent the points of the multi object position. The camera perspective problem is transformed into the control point problem under the camera coordinate system, and the coordinate points under the camera coordinate system are represented by the marked coordinate points to build the imaging model. The imaging model is corrected by using the model parameters of camera pinhole imaging, and the image coordinates are obtained. The camera internal parameters can be obtained by using the least square method, thus the unknown vector can be obtained. Identify the target points marked on the ground. If the number of points can not be completely recognized, the target position information can be obtained by identifying several of the points, so as to achieve multi-target visual positioning.

3.2   Error Motion Tracking Based on Multi Eye Machine Vision

Select appropriate instruments to establish a fast, accurate and effective imaging data entry and analysis system. When designing computer aided devices, it is necessary to make specific analysis on camera, special image processing system, lighting device and other components. The structure of the applied machine vision inspection system is shown in Fig. 4.

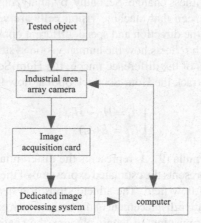

Fig. 4. Schematic diagram of application of multi eye machine vision inspection system

On the basis of athletes' wrong actions, the tracking process of wrong technical actions is designed by calculating athletes' wrong action descriptors. Multi eye machine vision technology is based on the characteristics of optical flow. The optical flow information set is established by the time displacement of each pixel, which requires high accuracy of optical flow. In order to transform the optical flow vector of moving video into a vector field, and then form a movable spatial distribution relationship, especially the optical flow field needs to be analyzed. Because optical flow can only reflect the motion information in the tennis player's foreground image when tracking the image, the background in the tracking image will affect the calculation of optical flow. Therefore, the background must be cleaned up first. Not only the uniformity of background color should be considered, but also the global foreground image of athletes should be obtained after the region growth algorithm is processed based on the Gaussian mixture model.

The inter frame difference method is used to eliminate the background color centered on athletes. By analyzing the gray difference of the corresponding pixels in the image, and then using double thresholds to select the gray difference in the image, the frame difference method is obtained by performing the difference operation on the adjacent frames in the image. When tracking athletes' wrong actions, there will be significant brightness difference between frames. Based on this, the absolute value of brightness difference between two frames can be expressed by the following formula:

$$\Delta G(a, b) = |G(a, b, t) - G(a, b, t - 1)| \tag{8}$$

In formula (8), $G(a, b, t)$ represents the grayscale value of image pixels at time t. According to the definition of the camera, the user can manually adjust the image sequence of the error action, and then estimate the length of the light field. First, according to the change degree of the camera's flash and intensity, the brightness of the error action image is tracked in real time, which will cause errors in the optical flow calculation results. Therefore, image difference is used to distinguish brightness and eliminate the influence caused by brightness change. Secondly, by analyzing the theory of biological vision system, it can be seen that machine vision cells are very sensitive to the edge movement of objects. In the direction and speed, different optical flows are formed due to different images, which reflects how the human vision system affects the changes of optical flow. On the basis of the difference image, the Hom-Schunck algorithm is used to estimate how athletes track the Hom-Schunck optical flow field, as follows:

$$\begin{cases} A_i = H_i - H_{i-1} \\ O_i = E(A_i) \end{cases} \tag{9}$$

In formula (9), In formula (9), A_i represents the differential image of tracking error action H_i and H_{i-1}; E represents the estimated expression of the Horn Chunck algorithm; O_i represents the optical flow field. The athlete's position in the adjusted error action image is related to the relative displacement of the body, which exists in the corresponding image area. For different postures of tennis players, the spatial distribution of optical flow field is different.

In the machine vision system, the binary image of a pixel in the image can be obtained by comparing the pixels between the correct action and the wrong action of the athlete, which can be expressed as:

$$I(a, b) = \begin{cases} 1 & \Delta G(a, b) < \xi \\ 0 & \Delta G(a, b) \geq \xi \end{cases} \tag{10}$$

In formula (10), ξ is the threshold for distinguishing between correct and incorrect actions by athletes. When the calculation result in the above formula is 0, it indicates that the difference of one pixel gray level in the two images is the difference data represented by the correct action of the athlete; When the calculation result in the above formula is 1, it indicates that the difference of one pixel gray level in the two images is the difference data represented by the athlete's wrong action. Between two adjacent frames, due to the environment, lighting and other factors, it is inevitable that there will be a pixel area with a calculation result of 1. At this time, it is necessary to determine whether this area is the area where the athlete's wrong action is located. In the machine vision system, the region screening method is used to select the pixel connected area with the operation value of 1 to judge whether this area is the wrong action area of athletes, so as to effectively screen the wrong action area of athletes and eliminate other interference.

According to the kernel density estimation and grid histogram of the error motion image, the optical flow histogram is collected as the motion descriptor of the athlete during the movement. For a given optical vector of the optical flow field coordinates, calculate the amplitude $\eta(o)$ and direction angle $\theta(o)$ of the athlete's erroneous action, with the formula:

$$\begin{cases} \eta(o) = \sqrt{\sigma_x^2(o) + \sigma_y^2(o)} \\ \theta(o) = \arctan \frac{\sigma_x(o)}{\sigma_y(o)} \end{cases} \tag{11}$$

In formula (11), $\sigma_x(o)$ and $\sigma_y(o)$ represent the horizontal and vertical components of the optical vector at the given optical flow field coordinate o. In conclusion, on the basis of multi eye machine vision technology, error action features are extracted, and error action descriptors are determined by tracking and adjusting error actions, so that error action tracking is realized.

4 Experiment

4.1 Establishment of Experimental Platform

In order to make full use of the convenience of mobile devices and the power of computer GPU floating point computing capability, the system adopts the C/S architecture mode, which is a typical distributed architecture. The system structure is shown in Fig. 5.

It can be seen from Fig. 5 that it mainly includes three parts, namely, the client, the network transport layer and the server.

The client refers to the mobile phone, which is mainly an athlete's long-term tracking system APP. First, you need to customize an IDL file, which is an essential work with the Thrif framework. First, define a structure to encapsulate the image frame of the video sequence, and then define two services. One is the init service to initialize the tracking algorithm of the server, and transfer the first frame (template image) and the initial position of the tracking target to the tracking algorithm for operation. The other is the tracking service. It is used to transfer the subsequent image frames (search images) to the tracking algorithm for target position prediction. Next, you can use the code generation engine of the Thrift framework to generate Java code, and write the corresponding

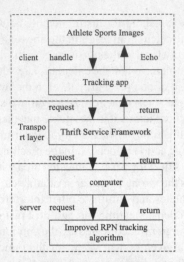

Fig. 5. Experimental Platform

operation interface and logic processing to build an Android client. It should be noted that the init service and tracking service are implemented on the server, and the client only borrows the injection interface. In addition, since the use of Thrift framework needs to consider the problem of network transmission delay, the video captured at the client adopts a caching strategy, which converts the video into a frame image, and then transmits it to the server for processing, and finally displays the results on the Android screen. Although this will cause the results displayed on the screen to be slightly slower than the actual shooting, it can effectively avoid the impact of network delay.

The network transport layer is implemented using TCP transport protocol and has been encapsulated by the Thrift framework.

The server of the server refers to the computer. Through the IDL file defined in the figure, the corresponding Python code can be generated. Then the improved SiamRPN algorithm can be written into the corresponding inti service and tracking service.

4.2 Experimental Evaluation Criteria

In order to quantitatively evaluate the athlete's error action tracking method based on multi eye machine vision, the recall rate index and accuracy rate index are introduced to determine the recognition ability of each action in the error action. The calculation method for recall index R and accuracy index P is:

$$\begin{cases} R = \frac{m_1}{m_1+m_2} \times 100\% \\ P = \frac{m_1}{m_1+m_3} \times 100\% \end{cases} \tag{12}$$

In formula (12), m_1, m_2, and m_3 respectively represent the number of correctly identified erroneous actions, the number of unrecognized erroneous actions, and the number of incorrectly identified erroneous actions.

4.3   Analysis of Experimental Data

Taking the high jumper as an example, the correct high jump action is shown in Fig. 6.

Fig. 6. Example of correct high jump

The correct action tracking curve and error action tracking curve corresponding to Fig. 6 are shown in Fig. 7.

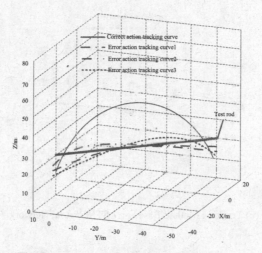

Fig. 7. Correct and error action tracking curve

It can be seen from Fig. 7 that the correct action tracking curve shows that the athlete successfully skips the test bar, while the wrong action tracking curve shows that the athlete fails to skip the test bar. The first type will hit the bar during the descent, the second type will hit the bar directly because of the short approach distance, and the third type will hit the bar because of the poor posture during the ascent.

4.4   Test Result

Use the tracking method based on the stacked short-term memory network, the tracking method based on LSTM neural network and the tracking method based on multi eye machine vision to compare and analyze whether the error action tracking curve is consistent with the curve connected to the experimental data, as shown in Fig. 8.

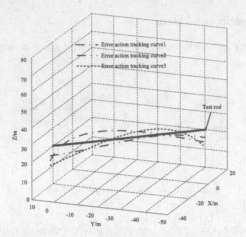

(a) Tracking Method Based on Stacked Long term and Short term Memory Network

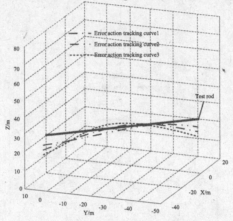

(b) Tracking Method Based on LSTM Neural Network

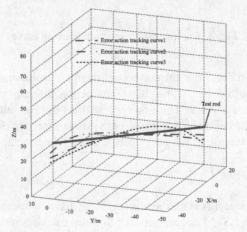

(c) Tracking method based on multi eye machine vision

Fig. 8. Error action tracking curve analysis of different methods

As shown in Fig. 8, using the tracking method based on the stacked short-term memory network and the tracking method based on the LSTM neural network, the three error action tracking curves are inconsistent with the actual curves, while using the tracking method based on multi eye machine vision, the three error action tracking curves are consistent with the actual curves.

In order to further verify the high tracking accuracy of the research method, the tracking effects of the three methods are compared again, as shown in Fig. 9.

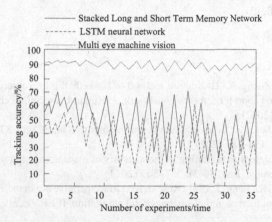

Fig. 9. Comparison and analysis of tracking effects of different methods

It can be seen from Fig. 9 that the maximum tracking accuracy of the tracking method based on the stacked short-term memory network is 71%, the maximum tracking accuracy of the tracking method based on LSTM neural network is 66%, and the maximum tracking accuracy of the tracking method based on multi eye machine vision is 93%.

5 Conclusion

With the continuous development and progress of society, people's attention to multi eye machine vision is increasing, and the research efforts of various institutions in this area have also been strengthened. Many research results have also been applied to our lives, and are widely used in entertainment, production, military and other aspects. This paper proposes a method of tracking athletes' wrong actions based on multi eye machine vision. After image recognition and analysis, the recognition and tracking of athletes' wrong actions are completed. The experimental results show that the research method can effectively track the wrong actions of athletes, and the identification parameters of foul actions are large, which has certain practical significance.

The outlook for future research is that with the continuous development of multi eye machine vision technology, more application scenarios will emerge. On the one hand, multi eye machine vision can play an important role in the field of sports training, helping coaches and athletes more accurately identify and correct incorrect movements, and improving training effectiveness. On the other hand, multi eye machine vision can also be applied in the medical field, helping rehabilitation training and treatment by monitoring and analyzing the patient's movement process. In addition, in the field of intelligent security, multi eye machine vision can provide more comprehensive and accurate monitoring and recognition functions, improving security and early warning capabilities. With further improvements in algorithms and hardware, multi eye machine vision technology will become more intelligent, efficient, and able to adapt to more complex environments and task requirements. In short, the research and application prospects of multi eye machine vision in the future are broad and will have a positive impact on various fields.

References

1. Niu, C., Lu, D., Zheng, Y.: Evaluation method of basketball player micro motions based on stackable long and short term memory network. J. Hunan Univ. Sci. Technol. Nat. Sci. Edn. **37**(02), 95–103 (2022)
2. Yang, S., Yang, J., Li, Z., et al.: Human action recognition based on LSTM neural network. J. Graph. **42**(02), 174–181 (2021)
3. Tao, L.: Application of data mining in the analysis of martial arts athlete competition skills and tactics. J. Healthc. Eng. **36**(3), 557–563 (2021)
4. Zhang, Y., Wang, K., Jiang, J., et al.: Research on intraoperative organ motion tracking method based on fusion of inertial and electromagnetic navigation. IEEE Access **51**(19), 4881–4889 (2021)
5. Wang, Y.: Real-time collection method of athletes' abnormal training data based on machine learning. Mob. Inf. Syst. **12**(3), 1–11 (2021)
6. Gu, K., Li, Y., You, X., et al.: Location tracking scanning method based on multi-focus in confocal coordinate measurement system. Precis. Eng. **32**(9), 170–177 (2021)
7. Liu, Y., Dong, H., Wang, L.: Trampoline motion decomposition method based on deep learning image recognition. Sci. Program. **114**(9), 1–8 (2021)
8. Guo, Y., Liu, Z., Luo, H., et al.: Multi-person multi-camera tracking for live stream videos based on improved motion model and matching cascade. Neurocomputing **492**(12), 561–571 (2022)
9. Wu, Z.: Human motion tracking algorithm based on image segmentation algorithm and Kinect depth information. Math. Probl. Eng. **41**(8), 64–69 (2021)
10. Zhao, J., Zhang, D.: Simulation of human motion information capture in time-space domain based on virtual reality. Comput. Simul. **38**(08), 391–395 (2021)
11. Pan, Y.L.: Run-up track tracking method of 110-meter hurdle athletes based on meanshift algorithm. J. Changchun Univ. **32**(10), 15–19 (2022)
12. Yang, J.: A sensor-based player tracking method. Microcomput. Appl. **39**(03), 12–16 (2023)
13. Wang, Y., Fang, W.C., Ma, J.W.: Posture segmentation based comprehensive assessment of actions. J. Signal Process. **38**(02), 300–308 (2022)
14. Ren, Y., Luo, J.T., Liang, X.P.: Algorithm for detecting occluded basketball players based on adaptive keypoint heatmap. J. Comput.-Aided Des. Comput. Graph. **33**(09), 1450–1456 (2021)
15. Luo, S., Qin, L.R.: Detection of basketball video events and key roles based on attention model. Comput. Appl. Softw. **38**(01), 186–191 (2021)

Research on Real Time Tracking Method of Multiple Moving Objects Based on Machine Vision

Yuan Wang(✉)

Wuhan Institute of Design and Sciences, Wuhan 430205, China
yinwar822172@163.com

Abstract. In the process of real-time tracking of multiple moving targets, there is a big gap between the tracking effect and the ideal effect due to the influence of the objective environment state. Therefore, a real-time tracking method of multiple moving targets based on machine vision is proposed. A machine vision acquisition unit including lighting device, camera, image acquisition card, processing system and control actuator is constructed. At the same time, in order to avoid the impact of outdoor extreme weather on image acquisition effect, MF-DSC03 with water-proof function and automatic light compensation function is used as the camera head device. In the phase of real-time tracking of multiple moving targets, based on HSV color features and apparent features in real-time video image data of multiple moving targets captured by machine vision, iterative tracking of the state of multiple moving targets is realized by means of KM bipartite graph matching algorithm. In the test results, the design method outperforms the control group in multi-target tracking performance on different data sets, showing an ideal tracking effect.

Keywords: Machine Vision · Multiple Moving Targets · Real-Time Tracking · Machine Vision Acquisition Unit · Hsv Color Characteristics · Apparent Characteristics · Km Bipartite Graph Matching Algorithm

1 Introduction

Computer vision is an important research direction in the field of artificial intelligence, which aims to study how to make computers perceive, analyze and process the real world intelligently like human vision systems. The camera is easy to install and use as a relatively inexpensive environment sensing sensor [1]. Therefore, various computer vision algorithms based on images and videos have already penetrated into the daily life of the public, such as face recognition technology used in access control and security inspection, intelligent monitoring in traffic field, visual navigation in automatic driving field, etc. [2]. Video target tracking technology, as one of the basic and important research directions in the field of computer vision, has always been the focus of researchers [3]. Generalized target tracking usually includes single object tracking (SOT) and multiple object tracking (MOT) [4]. Video target tracking focuses on the ability to locate

L. Yun et al. (Eds.): ADHIP 2023, LNICST 550, pp. 165–181, 2024.
https://doi.org/10.1007/978-3-031-50552-2_11

and estimate the size of an object of interest in a frame for a long time in subsequent video frames [5]. Usually, multi-target tracking is limited to scenes with known target categories, such as visual tracking of multiple pedestrians and vehicles [6]. In order to distinguish multiple targets, each target is usually marked with a digital ID representing a unique identity. Therefore, in addition to tracking the location and scale information of the target [7], simultaneous movement of multiple targets should also ensure that the same target remains fixed unique digital identification. At present, the research results of multi-target tracking methods show good tracking performance in some simple tracking scenarios, and there is no big gap [8]. As the difficulty of the scene increases, various evaluation indicators of tracking performance begin to decline in varying degrees[9]due to the different focus of different research results. The difficulty of tracking the scene mainly refers to the irregular movement of the target, the complexity of the target and the background, the similarity of the target and the interaction between the targets, etc. These scene difficulties are mainly manifested as occlusion problems in the multi target tracking research. Occlusion is the main cause of target loss during tracking [10]. Object occlusion can be divided into two situations: being occluded by background objects and being occluded by other objects. The occluded area will affect the probability of target loss to varying degrees. The situation that the target no longer appears in the scene after being occluded is generally classified as the situation that the target disappears, which is not within the scope of the occlusion problem. The occlusion problem focuses on the situation that the target reappears after being occluded. The occlusion phenomenon does not happen instantaneously but has a process, which can be divided into three stages; The first stage is the process of the target being blocked by obstacles, and the target gradually disappears from the image; The second stage is that the target is completely blocked by obstacles, at this time, the target is completely invisible in the image; The third stage is the process of the target gradually moving out of the obstacle, and the target information reappears in the image, which is also called target reappearance. The longer the whole process takes in the second stage, the longer the effective information of the target is lost, and the harder it is to find the target when the target reappears. At present, Most multi-target tracking algorithms do not have good methods to deal with the situation that the target is occluded for a long time. How to effectively deal with the occlusion problem will be the purpose of scholars' continuous research.

Therefore, this paper proposes a real-time tracking method for multiple moving targets based on machine vision. By introducing machine vision, this method collects the data of multiple moving targets, constructs the video image data acquisition process of moving targets, and the hardware used in machine vision, etc., and realizes target tracking through the combination of HSV color histogram and KM binary graph matching algorithm. Through the comparison test, the practical application effect of the designed tracking method is analyzed and verified, and the method has better tracking effect.

2 Design of Real-Time Tracking Method for Multiple Moving Targets

2.1 Data Acquisition Based on Machine Vision

Machine vision [11, 12] simulates and searches for human vision rules through computers, so as to input images into the image processing process. In order to meet the different functional requirements in different fields, understanding the user's performance requirements for each component of data collection is the key to designing a good machine vision product. And the application of machine vision is because the method has higher automation, efficiency, stability, diversity and flexibility, which can process a large number of image data in a short time. The main structure of machine vision acquisition includes: lighting device, camera, image acquisition card, processing system and control actuator. For a machine vision system, there can be several input signals, which can be obtained from different perspectives, different times or different directions. The obtained 3D scene projection signal is input into the system. Usually, the input signal represents the spatial relationship of objects, such as surface spatial structure, surface shape characteristics, texture characteristics, color characteristics and other information. The cooperation between the modules of the machine vision system directly affects the function of the whole system. Any problem in any module will directly affect the performance of the next module or even the whole system, and eventually lead to deviation or error in the system results. Obtaining high-quality image information is the premise of machine vision system. However, in order to obtain ideal image quality, we must start from the selection of acquisition methods, lens selection and other aspects, then input the obtained image information into the processing platform, and cooperate with the module controller to achieve the higher-level functions of machine vision system.

In order to track pedestrians in real life, this paper provides the remote monitoring data adopts the camera for image shooting, and processes the image data information collected by the camera, so as to achieve the goal of target tracking. The flow chart of image acquisition used in this study is shown in Fig. 1.

The key step in the target tracking process is to use the camera to obtain the image, obtain the external environment information, and further process the acquired image. In order to achieve better real-time image acquisition, the main components needed include: light source, light source controller and lens. The display displays the image processing effect in real time, which is convenient for human observation. The image acquisition module usually uses CMOS or CCD cameras to convert the collected image information into digital signals and transmit them to the software platform with processing functions. According to the specific needs of different environments, different processing algorithms are used to achieve the analysis and processing of targets, extract the target area of interest, and make real-time judgments on the processing results, Then control relevant equipment to respond, achieve the purpose of image processing module, and provide basis for more advanced behavior.

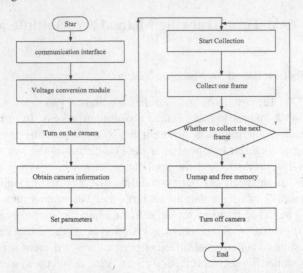

Fig. 1. Video image data acquisition process based on machine vision

The camera is the core of the image acquisition system. This paper selects a camera with waterproof function and automatic light compensation function, which can still achieve the purpose of image acquisition in extreme weather outdoor. The MF-DSC03 series serial camera used in this research is a highly integrated embedded digital camera. The camera has 2 million pixels and dedicated image compression technology. The DSP microprocessor is used for image compression processing. The image quality is clear, the color is lifelike, and it has the function of infrared night vision light compensation, Its image sensor adopts the world's best 0V series high-performance CMOS sensor, with progressive scanning function, which has high practical value in traffic supervision, intelligent agricultural production and other aspects. The output image format of MF-DSC03 series products can be directly converted to JPEG format.The data is transmitted through RS232/RS485/TL, and the electronic shutter is used for shooting. The MF-DSC03 series serial port camera heads are best used in places that require high image quality and need to transmit images. The main performance parameters of MF-DSC03 series serial camera are shown in Table 1.

On this basis, the real-time video image data of multiple moving objects is collected using machine vision, which provides the basis for the follow-up target tracking.

2.2 Real Time Tracking of Multiple Moving Targets

In most tracking scenes [13, 14], the color information of the target is the most recognizable appearance description, so this paper chooses to extract HSV color histogram statistical features of the target to represent its color information. The reason for choosing the HSV color histogram is that the HSV space can express the light and shade, tone, and vividness of the color very intuitively, so as to facilitate the contrast between colors. However, in a few tracking scenes, there will be target objects with highly similar appearance, whose color characteristics are not enough to distinguish each other's

Table 1. Parameters of MF-DSC03 Series Serial Camera

S/N	function	model
1	Image sensor type	CMOS 1/4 inch
2	Camera pixels	2 million
3	Pixel size	2.2umX22um
4	Output format	JPEG format
5	Balance mode	automatic
6	Exposure mode	automatic
7	Gain mode	automatic
8	Signal-to-noise ratio	40 dB
9	dynamic range	50 dB
10	Maximum analog gain	16 dB
11	Frame rate	UXGASXGA: 15fps SVGA: 30 fps CIF: 60fps
12	viewing angle	90 degrees (120 or other angles are optional)
13	Distance range	0-20m (adjust according to specific requirements)
14	Serial port speed	Between 15000 and 120000
15	Operating current value	≤ 100 mA (turn off the infrared device) ≤ 200 mA (turn on the infrared device)
16	Operating voltage value	DC 5V

identities, and sometimes the deformation of targets will lead to differences in color histogram statistics. Considering that in the Re ID field, the convolutional network can extract differentiated depth information from the appearance of targets, It can make similar targets show differences at the depth semantic feature level. Therefore, the fusion of depth apparent features enhances the feature representation of targets.

To sum up, the motion information generated by the extended Kalman filter for position prediction, the color information of the HSV color histogram [15] and the apparent information of the Re ID depth feature are fully considered, and the three feature information are fused to measure the similarity between the detection sequence and the tracking track sequence, so as to increase the accuracy of data association.

For HSV color feature measurement, HSV corresponds to hue (HueE (0, 360)), saturation (SaturationE (0, 1)), and brightness (ValueE (0, 1)) respectively, which is a color space constructed according to the intuitive features of image color, and can reflect the true color information of objects. Therefore, HSV color feature histogram features are used to describe the color information of detection and tracking tracks. The three channel colors in the HSV space are uniformly quantized. The specific uniform quantization method can be expressed as

$$
H \begin{cases} 0, H \in [0, 22.5) \\ 1, H \in [22.5, 45) \\ 2, H \in [45, 67.5) \\ \dots \\ 14, H \in [315, 337.5) \\ 15, H \in [337.5, 360) \end{cases} \tag{1}
$$

$$
S \begin{cases} 0, S \in [0, 0.25) \\ 1, S \in [0.25, 0.5) \\ 2, S \in [0.5, 0.75) \\ 3, S \in [0.75, 1) \end{cases} \tag{2}
$$

$$
V \begin{cases} 0, V \in [0, 0.25) \\ 1, V \in [0.25, 0.5) \\ 2, V \in [0.5, 0.75) \\ 3, V \in [0.75, 1) \end{cases} \tag{3}
$$

On this basis, the quantized histogram eigenvalues of real-time video image data of multiple moving targets are used V_{HSV} Represents, which can be expressed as

$$
\begin{aligned} V_{HSV} &= 16H + 4S + V \\ V_{HSV} &\in (0, 255) \end{aligned} \tag{4}
$$

For color characteristics f_{color} express, f_{color}. It represents the 256 dimensional color histogram feature vector of a target. The chi square distance is used to measure the color difference, which can be expressed as

$$
d_{color}(i, j) = \sum \frac{[f_{color(i)}(n) - f_{color(j)}(n)]^2}{f_{color(i)}(n) + f_{color(j)}(n)} \tag{5}
$$

where, $d_{color}(i, j)$ Represents the chi square distance of colors in real-time video image data of multiple moving targets, $f_{color(i)}(n)$ Record as detection target i Color characteristics, $f_{color(j)}(n)$ Record as track j Color characteristics of.

For the apparent feature measurement of real-time video image data of multiple moving objects, the specific implementation method is.

Initialize the target track with identity information Q Set to

$$
Q = \{Q_1, Q_2, Q_3 \dots Q_n\} \tag{6}
$$

The track information is transmitted frame by frame, and then the detected target sequence is measured in the next frame according to its apparent characteristics G Expressed as

$$
G = \{G_1, G_2, G_3 \dots G_n\} \tag{7}
$$

Use it with Q According to the measurement results, the target is re identified and confirmed. For this purpose, this article uses f_{reid} To describe the apparent characteristics of a target, f_{reid} Specifically, it refers to the 128 dimensional apparent feature vector extracted. Assuming that there are N detection objects in the current frame, their corresponding characteristic matrix F_D Can be expressed as

$$F_D = \{f_{reid(1)}, f_{reid(2)}, f_{reid(3)}, ..., f_{reid(N)}\} \tag{8}$$

For M confirmed tracking tracks, each confirmed associated track T (j) stores the composition set of its previous 50 frame feature vectors $F_{confirm}^{T(j)}$ Can be expressed as

$$F_{confirm}^{T(j)} = \{f_{reid(1)}^{T(j)}, f_{reid(2)}^{T(j)}, f_{reid(3)}^{T(j)}, ..., f_{reid(N)}^{T(j)}\} \tag{9}$$

Here, the number of temporary features is adjusted by setting 50 super parameters. By calculating the number of i Apparent characteristics of detected objects $f_{reid(j)}$ And the j 50 frames of tracks $F_{confirm}^{T(j)}$ The minimum cosine distance between is used to measure the apparent similarity. The specific calculation method can be expressed as

$$d_{reid}(i,j) = \min\{1 - f_{reid}^{T(j)} F_{confirm}^{T(j)}\} \tag{10}$$

Among them, $d_{reid}(i,j)$ Represent each feature difference parameter of the tracking track, calculate the cosine distance between them and the apparent features of N detection objects in the current frame, and then take the minimum value as the calculation value between the track and the detection result, and the cosine cost matrix can be expressed as

$$\cos t_{\cos(M*N)} = \begin{bmatrix} d_{reid}(1,1)......d_{reid}(1,N) \\ d_{reid}(2,1)......d_{reid}(3,N) \\ .... \\ d_{reid}(M,1)......d_{reid}(M,N) \end{bmatrix} \tag{11}$$

Among them, $\cos t_{\cos(M*N)}$ The cosine cost matrix representing the real-time video image data of multiple moving targets.

On this basis, because the number of targets in the tracking scene is constantly changing, not all of the tracks in each frame can form a one-to-one correspondence relationship with the detection data after association, there must be unmatched tracks and detection targets [16–18], so it is necessary to divide the tracks and detection states. The KM bipartite graph matching algorithm is introduced, because the algorithm is more efficient, comparable to the cost flow, much simpler than the cost flow, and the idea is clearer, and solves the problem of maximum weight matching, which is suitable for the state division of trajectory and detection. After calculation by KM bipartite graph matching algorithm, set the two states of matched detection and unmatched detection, and set the two states of unconfirmed and confirmed for all tracks. The confirmed state refers to the track that has completed the correct matching. In the current frame, the confirmed state is divided into unmatched track and matched track, and the unconfirmed track is initialized by the unmatched detection target. The unconfirmed track is set to reduce the chance of matching and prevent the detection target from not being associated with tracks of the same identity. The specific method is to set the matching threshold Hits = 2 frames (too many frames will affect real-time), that is, the KM algorithm completes the matching of two consecutive frames before it is allowed to initialize to the confirmed track. Because the target is blocked or the detector misses detection, the tracks in the confirmed state will appear unmatched tracks [19, 20]. In this case, set the lifetime Age and consecutive lost frame count Count for the confirmation state track. Where Age = 100 (if the value is too high, the garbage track will be cached too much) indicates the upper limit of the number of consecutive lost frames. When 0SCountSAge, the track is valid. In the matching process, count + 1 iterative correlation is performed to match the tracks with fewer lost frames in priority, while ensuring the effectiveness of the lost tracks in the survival period. The specific implementation process of the update algorithm is divided into the following steps:

(1) Set the variable Count = 0 for the number of consecutive lost frames and the constant Age = 100 for the lifetime;
(2) At time t, n detection objects are output by machine vision and numbered $d \in \{0, 1,2,3,\ldots n\}$, d contains unmatched detection objects and matched detection objects in the previous frame;
(3) At time t, the m tracks predicted by the extended Kalman filter in this paper are numbered $k \in \{0, 1,2, 3, m\}$, and each track has a Count variable to record the number of consecutive lost frames of tracks, and k contains unmatched tracks and matched tracks whose Count is less than Age times;
(4) Use the weighted multi feature fusion algorithm proposed in Sect. 4.23 of this paper to measure the similarity between d and k, and output the cost matrix $\cos t_{\cos(M*N)}$ The KM algorithm is used to perform Count + 1 iteration matching association. This process starts from the track with Count = 0 and iterates to the maximum of Count, that is, the stable tracking track is first associated;
(5) After iterative correlation, the unmatched track Count value is increased by 1, and the matched track Count is set to 0, unmatched

The detection target of is initialized as an unconfirmed track, and the track whose Count is greater than Age is deleted.

So far, the status update of all tracks in one frame has been completed, realizing real-time tracking of moving objects.

3 Test Analysis

3.1 Test Data Set Preparation

In order to verify the effectiveness of multi-target tracking algorithm based on attention mechanism in improving the accuracy of tracking algorithm, this chapter conducts experiments on MOT17, MOT16 and MOT15 datasets respectively, and uses MOTA, MOTP, IDF1, MT, ML, FP, FN and IDS to evaluate and analyze.

MOT Challenge is a public platform that can upload and publish multi-target tracking research results, and has the largest pedestrian tracking data set. The platform is designed to evaluate the detection performance and tracking performance of multi pedestrian in complex environments, mainly including MOT15, MOT16 and MOT17 data sets, which are introduced below. Table 2 shows the detailed attribute information of the dataset.

Table 2. Basic Information Statistics of Test Data Set

data set	Video source	length	Number of tracks	FPS	Camera condition	launch a pilot project
MOT15	TUD-Crossing	201	13	25	static	level
	PETS2009-S21.2	436	42	7	static	high
	ETH-C rossi ng	219	26	14	move	low
	ADL -R undle-l	500	32	30	move	level
	KITTI-16	209	17	10	static	level
MOT16	MOT1 6–01	450	23	30	static	level
	MOT1 6–03	1500	148	30	move	high
	MOT16–06	1194	221	14	move	high
	MOT1 6–12	900	86	30	static	level
MOT17	MOT1 7–01	450	24	30	static	level
	MOT1 7–03	1500	148	30	move	high
	MOT1 7–06	1194	222	14	move	level

The basic situation of three test data sets is analyzed.

(1) MOT15: The data set disclosed by MOT Challenge in 2015, in which video sequences are shot by static or mobile cameras under unlimited conditions. The data set contains 22 video sequences, 11 of which are test sets and 11 are training sets. There are various scenes. The dataset has 11283 frames in total, including 1221 different pedestrians and 101345 pedestrian detection frames.

(2) MOTI6: The data set released by MOT Challenge in 2016 includes 7 training sets and 7 test sets, totaling 14 video sets. Compared with MOTI5 dataset, its annotation is more accurate, the scene is more complex, and the number of pedestrians is more, including 11235 frames, including 1342 pedestrians and 292733 pedestrian detection frames. The official test results are the results detected by DPM detector.

(3) MOT17: The dataset released by MOT Challenge in 2017 has the same video sequence as MOT16. The only difference is that the targets in the dataset are detected by three different detectors (Faster R-CNN, DPM, SDP), so the detection results are more accurate.

On this basis, multi moving target tracking test is carried out.

3.2 Multi Target Tracking Evaluation Indicator Setting

In order to evaluate the performance of multi-target tracking algorithm fairly and accurately, Table 3 lists the most commonly used multi-target tracking evaluation indicators.

Table 3. Multi target tracking evaluation indicators

S/N	Measure Name	Expected score	sketch
1	MOTA↑	100%	Multi target tracking accuracy
2	MOTP↑	100%	Multi target tracking accuracy
3	IDF1↑	100%	Accuracy of identity representation
4	MT↑	100%	Most tracked targets
5	ML↓	0%	The least lost goal
6	FP↓	0	Number of unmatched tracks
7	FN↓	0	No matched real target
8	IDS↓	0	Total number of identity switches

Carry out specific analysis on the multi-target tracking evaluation indicators shown in Table 3.

(1) MOTA: Accuracy of multi-target tracking. This indicator is used to determine the number of targets and the cumulative error of statistical tracking. It is mainly related to three factors: FP (number of false detections), FN (number of missed detections), and IDSW (number of identity information handovers). The fewer false positives, false positives, and identity switches in the tracking process, the higher the tracking accuracy of the model. The calculation formula can be expressed as

$$MOTA = 1 - \sum \frac{(m_t + f_{pt} + mme_t)}{\sum g_t} \tag{12}$$

Among them, m_t Section t Number of missed frames, f_{pt} For t Number of false detections of frames, mme_t Is the number of mismatches IDSW, that is, the t. The number of targets with identity switching in the frame, g_t For t The number of real targets in the frame.

(2) MOTP: Accuracy of multi-target tracking model. This indicator is used to evaluate the precision of the predicted target box position

Accuracy is generally defined by calculating the coincidence rate between the prediction box and the real dimension box. The calculation formula can be expressed as

$$MOTP = \frac{\sum d_i^t}{\sum c_t} \tag{13}$$

Among them. c_t Represents the t. The number of successful matches between the real target and the predicted target in the frame, d_i^t It indicates the Euclidean distance between target matching pairs, i.e. error.

(3) MT: indicates the proportion of most tracked target tracks, that is, when the tracks tracked account for more than 80% of the real track length of the target, most of the target is tracked.

(4) ML: indicates the proportion of most untracked target tracks, that is, when the tracks tracked by the target account for less than 20% of the real track length of the target, most of the target tracks are missing.

(5) FP: indicates the number of false checks, that is, the number of tracks not matched in the current frame.

(6) FN: indicates the number of missed checks, that is, the number of true annotations on the current frame that cannot be matched.

(7) IDS: The number of times the trace changed the identity information.

(8) IDE1: measures the accuracy of identity in tracking.

Based on the multiple target tracking evaluation indicators set above, the specific tracking effect is analyzed.

3.3 Systematic Testing

In order to ensure the accuracy of the experiment, it is necessary to ensure the normal operation of each module designed in this paper. Therefore, systematic testing is carried out to ensure the effectiveness of the module and the accuracy of the experiment. The test results are shown in Table 4.

Table 4. Systematic test results

Function name	Status
Moving target image video acquisition	Normal
Data transmission	Normal
Target image display	Normal
Data compression	Normal
Target feature extraction	Normal
Target matching	Normal
Target tracking	Normal

According to Table 4, the functions of each module of the system designed in this paper can run normally, and it has the basic feasibility of experimental analysis of target tracking.

3.4 Test Results and Analysis

In this section, on the open data sets of MOT17, MOT16 and MOT15, the attention mechanism based multi-target tracking method proposed in this chapter is compared with the reference [4] method and the reference [5] method. The reference [4] method is a moving target location method of moving image sequence based on the optimized particle filter hybrid tracking algorithm. The optimized particle filter hybrid tracking algorithm is introduced to achieve target tracking and positioning, while the reference [5] method is a fast moving and deformed target tracking method based on heterogeneous feature fusion, which realizes target tracking by integrating heterogeneous features. The experimental results of the three methods are shown in Tables 5, 6 and 7 below.

Table 5. Statistical Table of Comparison Experiment Results on MOT15 Data Set

evaluating indicator	Tracking method		
	Reference [4] method	Reference [5] method	Design method in this paper
MOTA	44.10	46.33	46.59
MOTP	77.50	53.40	79.10
IDF1	46.00	46.69	47.59
MT	17.20	19.00	18.31
ML	26.60	27.74	28.02
FP	6058	4627	4558
FN	26917	27065	26951
IDS	1347	1280	1304

It can be seen from Table 5 to Table 7 that on the MOT17 data set, the real-time tracking method for multiple moving targets proposed in this paper has increased by 2.58%, 1.61%, 1.24% and 1.53% on MOTA, MOTP, IDF1 and MT indicators respectively, and decreased by 1.23%, 3041 and 11471 on ML, FP and FN indicators respectively, compared with reference [4] method. On MOT16 dataset, the multi moving target real-time tracking method proposed in this paper has also greatly improved the evaluation indicators other than IDS. On the MOT15 dataset, the multi moving target real-time tracking method proposed in this paper is only higher than reference [4] method in ML and FN. This may be because MOT15 dataset is collected in an unconstrained environment, and the pedestrian in the video is too small, resulting in the loss of tracking targets. However, in general, the performance of the multi moving target real-time tracking method

Table 6. Statistical Table of Comparison Experiment Results on MOT16 Data Set

evaluating indicator	Tracking method		
	Reference [4] method	Reference [5] method	Design method in this paper
MOTA	54.80	56.86	56.95
MOTP	54.15	78.81	55.04
IDF1	53.40	55.14	55.51
MT	19.10	20.82	21.08
ML	37.00	34.91	34.91
FP	2955	2673	2625
FN	78765	75334	75217
IDS	645	647	644

Table 7. Statistical Table of Comparison Experiment Results on MOT15 Data Set

evaluating indicator	Tracking method		
	Reference [4] method	Reference [5] method	Design method in this paper
MOTA	53.70	56.11	56.28
MOTP	77.20	53.80	75.25
IDF1	53.80	54.15	55.04
MT	19.40	20.38	20.93
ML	36.60	35.50	35.37
FP	11731	8854	8690
FN	247447	237008	235976
IDS	1947	2065	1968

designed in this paper based on adaptive particle swarm optimization algorithm is significantly better than that of reference [4] method. In this paper, the tracking method is designed to reconstruct the features of adjacent frame targets through adaptive mechanism, so that the target features can ignore the similar features within the frame, enhance the feature expression ability of inter frame targets, and the correlation matrix obtained by relying on the reconstructed features is more reliable, thus improving the tracking accuracy of multi-target tracking.

On MOT17 dataset, the real-time tracking method for multiple moving targets proposed in this paper is 0.17% higher, 3.56% higher, 0.89% higher and 0.55% lower in MOTA, MOTP, IDF1 and MT than in reference [5] method, and 0.13% lower, 1032 and 97 lower in ML, FN and IDS respectively. On MOT16 data set, the real-time tracking method of multiple moving targets proposed in this paper is superior to reference [5] method in all indicators. On the MOT15 dataset, the multi moving target real-time tracking method proposed in this paper is superior to the reference [5] method in MOTA, MOTP, IDF1, FP, FN indicators, but the MT indicators are low and the ML indicators are high, which may be due to the unstable tracking results caused by the MOT15 dataset's own attributes. In general, the multi moving target real-time tracking method based on adaptive particle swarm optimization algorithm is superior to the reference [5] method in many indicators, which may be because reference [5] method heavily relies on the target dependency set manually, is sensitive to the setting of many super parameters, and fails to fully exploit the correlation between targets, However, this adaptive framework is introduced into the multi moving target real-time tracking method proposed in this paper. The adaptive mechanism is used to learn the dependency between targets in the frame, and the cross attention mechanism is used to learn the correlation between targets in the frame, so that high-quality target dependency can be automatically learned, and the tracking performance of multi target tracking is improved.

In order to more intuitively compare the tracking performance of the reference [5] method and the multi moving target real-time tracking method proposed in this paper, Fig. 2 shows some visualization results of the two algorithms in the two video sequences of MOT17 dataset 02 and 09.

It can be seen from Fig. 2 that Fig. 2 - (a) compares the tracking effect under the complex background with dim light. From the figure, it can be seen that in the 308th frame of MOT17–02 dataset, the real-time tracking method of multiple moving targets proposed in this paper can track the smaller target 37 compared with the reference [5] method.

Figure 2 - (b) compares the tracking effect when the target is blocked by a static object. It can be seen from the figure that at frame 391 of MOT17–09 dataset, the real-time tracking method of multiple moving targets proposed in this paper can track 29 targets blocked by buildings very well.

Therefore, combined with the visual test results, it can be further concluded that the multi moving target real-time tracking method proposed in this paper further improves the tracking accuracy compared with the reference [5] method.

Reference [5] method

This article designs a tracking method

(a) Comparison Diagram of Tracking Effect under Complex Background with Dim Light

Reference [5] method

This article designs a tracking method

(b) Comparison of tracking effect when the target is statically occluded

Fig. 2. Comparison of real-time tracking effects of multiple moving targets

4 Conclusion

Real time online tracking of multiple moving objects in video has always been a key task in computer vision, and plays an important role in camera based environment perception. For example, the practical application of multi-target tracking technology can be seen in scenes such as automatic driving, security and traffic monitoring. Since the development of online multi-target tracking technology, the technical target has always been how to maintain the trajectory of multiple objects of interest from beginning to end without loss. To achieve this goal, it is necessary to solve the problems of object of interest detection, motion trajectory prediction, detection and trajectory correlation, and object occlusion. In this paper, a real-time tracking method for multiple moving objects based on machine vision is proposed, which can accurately locate multiple specific objects in the video during the multi-target tracking phase, maintain the identity information of the target in complex scenes, and output complete target motion tracks. Different from single target tracking, the multi-target tracking designed in this paper not only realizes the purpose of accurately outputting the position of each target in the video frame, but also realizes the effective target matching between frames.

Acknowledgement. 2022 Hubei Provincial Department of Education Science Research Program Guiding Project: Application of Gesture Interaction Technology in Public Art Design in Smart Cities (B2022422).

References

1. Subrahmanyam, K., Kavitha, L.M., Rao, S.K.: Shifted Rayleigh filter: a novel estimation filtering algorithm for pervasive underwater passive target tracking for computation in 3D by bearing and elevation measurements. Int. J. Pervas. Comput. Commun. **18**(3), 272–287 (2022)
2. Wang, X., Xie, W., Liangqun, L.I.: Labeled Multi-Bernoulli Maneuvering target tracking algorithm via TSK iterative regression model. Chin. J. Electron. **31**(2), 227–239 (2022)
3. Dong, J., Li, Y., Guo, Q., Liang, X.: Through-wall moving target tracking algorithm in multipath using UWB radar. IEEE Geosci. Rem. Sens. Lett. **19**, 1–5 (2022). https://doi.org/10.1109/LGRS.2021.3050501
4. Guo, Y.: Moving target localization in sports image sequence based on optimized particle filter hybrid tracking algorithm. Complexity **2021**(7), 1–11 (2021)
5. Li, B., Jing, Q.: Fast moving and deformational target tracking approach based on heterogeneous features fusion. Trans. Inst. Meas. Control. **43**(3), 612–622 (2021)
6. Yang, J., Liu, X., Sun, J., Li, S.: Sampled-data robust visual servoing control for moving target tracking of an inertially stabilized platform with a measurement delay. Automatica **137**, 110105 (2022). https://doi.org/10.1016/j.automatica.2021.110105
7. Guo, C., You, S., Luo, Y., et al.: A fast moving target detection tracking and trajectory prediction system for binocular vision. Wuhan Univ. J. Nat. Sci. **26**(1), 12 (2021)
8. Xu, W.: Adaptive probability hypothesis density filter for multi-target tracking with unknown measurement noise statistics. Measure. Control London Inst. Measure. Control **54**(3–4), 002029402199280 (2021)
9. Lin, C., Shi, J., Lyu, Y., et al.: Over-flight and standoff tracking of a ground target with a fixed-wing unmanned aerial vehicle based on a unified sliding mode guidance law. Trans. Inst. Measure. Control. **44**(2), 410–423 (2022)

10. Zhou, C., Wang, Y., Xiong, R.H., et al.: Research on improvement of efficiency of radio monitoring and positioning based on UAV. Comput. Simul. **003**, 039 (2022)
11. Zhang, Y., Zhao, J., Han, H.: A 3D machine vision-enabled intelligent robot architecture. Mob. Inf. Syst. **2021**, 1–11 (2021). https://doi.org/10.1155/2021/6617286
12. Ahn, J., Kim, S.: Automated textile circuit generation using machine vision and embroidery technique. Textile Res. J. **92**(11–12), 1977–1986 (2022). https://doi.org/10.1177/004051752 21075062
13. Du, D.: Multiperson target dynamic tracking method for athlete training based on wireless body area network. Adv. Math. Phys. **2021**, 1–9 (2021). https://doi.org/10.1155/2021/228 7751
14. Wang, X., Xie, W., Li, L.: Labeled multi-Bernoulli Maneuvering target tracking algorithm via TSK iterative regression model. Chinese J. Electron. **31**(2), 227–239 (2022). https://doi.org/10.1049/cje.2020.00.156
15. Lv, C., Li, J., Kou, Q., et al.: Stereo matching algorithm based on hsv color space and improved census transform. Math. Probl. Eng. **2021**(32), 1–17 (2021)
16. Yang, D.: Target tracking algorithm based on adaptive scale detection learning. Complexity **2021**, 1–11 (2021). https://doi.org/10.1155/2021/9033912
17. Zhang, H., Cheng, L., Zhang, J., Huang, W., Liu, X., Yu, J.: Structural pixel-wise target attention for robust object tracking. Digital Signal Process. **117**, 103139 (2021). https://doi.org/10.1016/j.dsp.2021.103139
18. Liu, Y., Wang, X.: Mean shift fusion color histogram algorithm for nonrigid complex target tracking in sports video. Complexity **2021**, 1–11 (2021). https://doi.org/10.1155/2021/556 9637
19. Pirovano, L., Armellin, R., Siminski, J., Flohrer, T.: Differential algebra enabled multi-target tracking for too-short arcs. Acta Astronautica **182**, 310–324 (2021). https://doi.org/10.1016/j.actaastro.2021.02.023
20. Song, D., Gan, W., Yao, P., Zang, W., Zhang, Z., Qu, X.: Guidance and control of autonomous surface underwater vehicles for target tracking in ocean environment by deep reinforcement learning. Ocean Eng. **250**, 110947 (2022). https://doi.org/10.1016/j.oceaneng.2022.110947

A Method for Identity Feature Recognition in Wireless Visual Sensing Networks Based on Convolutional Neural Networks

Chenyang Li[✉] and Zhiyu Huang

Shenyang Institute of Technology, Shenyang 113122, China
liklu22@163.com

Abstract. Due to the problems of low recognition accuracy and long recognition time in traditional wireless visual sensing network identity feature recognition methods, a convolutional neural network-based wireless visual senscto the operation results, the global threshold method is used to obtain the binary image sequence and perform morphological processing. Based on the processing results, Extract target regions from video image sequences of wireless visual sensing networks, detect human targets, and construct a Softmax classifier using convolutional neural networks to classify human targets in video image sequences of wireless visual sensing networks, in order to identify identity features. The simulation results show that the proposed method has high accuracy and short recognition time for identity feature recognition in wireless visual sensing networks.

Keywords: Convolutional Neural Network · Wireless Visual Sensing Network · Identity Feature Recognition · Image Sequence · Mean Method

1 Introduction

The wireless vision sensor network is composed of multiple wireless nodes integrated with miniature vision sensors, which can transmit the acquired visual perception information to the Sink node (Sink node) through the cooperative way of multiple nodes, and then send it to the application server for subsequent processing and analysis. Wireless vision sensor networks not only have the advantages of traditional wireless sensor networks, such as self-organization, self-healing, flexible configuration, fast coverage and low-cost deployment, but also have the characteristics of the traditional vision application system with rich information, which can support a wider range of intelligent applications, such as traffic monitoring and traffic statistics, assisted living, public behavior analysis and modeling, and virtual reality. However, the resources of unlicensed frequency band suitable for wireless multi-hop transmission of visual information are limited, which limits the scale and performance of the network. Using wireless technology to access idle authorized frequency bands is one of the feasible ways to enhance the performance of wireless visual sensing networks and achieve large-scale applications[1]. The main challenge of wireless visual sensing networks is the randomness of available spectra

© ICST Institute for Computer Sciences, Social Informatics and Telecommunications Engineering 2024
Published by Springer Nature Switzerland AG 2024. All Rights Reserved
L. Yun et al. (Eds.): ADHIP 2023, LNICST 550, pp. 182–198, 2024.
https://doi.org/10.1007/978-3-031-50552-2_12

or channels. Due to the opportunistic access of wireless nodes to the idle authorized spectrum, their underlying link transmission capacity is dynamically changing. Under the traditional design principles of layered network protocols, the transmission of upper level visual information does not adaptively match changes in the transmission capacity of the lower level, thus unable to fully utilize the benefits brought by radio. This requires the use of cross layer design methods to enable upper layer applications to perceive potential transmission opportunities on lower layer links in real-time, in order to enhance the end-to-end service quality of visual information [2].

With the rapid development of information technology, wireless visual sensor network technology has been widely used in finance, e-commerce and other fields. The information of wireless visual sensor network is increasing rapidly, and the security of sensor network information has become a key issue in this field. How to effectively increase the security performance of wireless visual sensor network has become an urgent problem to be solved, and identity feature recognition is a necessary prerequisite to ensure the security of visual sensor network. How to effectively identify the user's identity and protect the security of visual sensor network information has been widely paid attention by experts and scholars in related fields. At the same time, there are also some good adaptive identity feature recognition algorithms [3]. Literature [4] proposed that the difference of footstep induced structural vibration signals in the walking process was used to identify personnel. Based on the energy threshold method, footstep events and non-footstep events were detected. A total of 16 footstep characteristic parameters of a single footstep event for different test personnel were compared and analyzed in the time domain and frequency domain. It is found that the parameter difference under different feature combinations can be used as the basis for identity recognition. In order to verify the effectiveness of the method, support vector machine (SVM) was used as a classification tool. With a test population of 10 people and 500 data samples, 16 foot feature parameters were selected with an average recognition rate of 79.21%. The Pearson correlation coefficient method was used to screen out 10 unrelated foot feature parameters with an average recognition rate of 91%, which was 11.79% higher than the average recognition rate using 16 foot feature parameters, We compared the impact of classification tools on the average recognition rate of 10 selected foot feature parameters under different SVM kernel functions, and found that the highest average recognition rate was 96% under linear kernel functions. The results indicate that an effective combination of foot feature parameters is suitable for identity recognition in small samples. However, the accuracy of the above methods for identity feature recognition is relatively low, resulting in poor recognition performance. Literature [5] uses BGN semi homomorphic encryption algorithm and Shamir secret sharing to design a threshold identity scheme based on biometric identification, which mainly uses BGN homomorphic encryption algorithm on bilinear pairs for data protection, uses a third-party authentication center for secret segmentation, and the server authenticates the user's identity in the ciphertext state to achieve threshold identity authentication. Literature [6] proposes an online detection and automatic identification technology for network video monitoring devices. Stateless scanning technology is used to carry out online detection of network terminal devices, extract BANNER and HTML page information from HTTP header information returned from specific ports of terminal devices, and construct Web identity features of devices through rough set attribute reduction. The cosine distance is used to calculate the similarity between the Web identity features of online devices and the sample of the known device signature database to realize the detection and identification of online devices. However, it takes a long time for the above two methods to recognize the identity features, resulting in low efficiency.

In response to the problems existing in the above methods, this paper proposes a wireless visual sensing network identity feature recognition method based on convolutional neural networks. Firstly, the architecture of wireless visual sensor networks is analyzed. Then, the background image of the application scene is processed by background subtraction to realize the target detection in the video image sequence. Using convolutional neural network, the Softmax classifier is constructed to classify the human body in the video image sequence of the wireless visual sensor network, and finally realize the recognition of identity features. Simulation experiments have verified that this method can quickly and accurately recognize the identity features of wireless visual sensing networks, laying a certain foundation for the safe operation of wireless visual sensing networks.

2   Wireless Visual Sensor Network Identity Feature Recognition Method

2.1   Analysis of Wireless Vision Sensor Network Architecture

The wireless vision sensor network system preloads part of AI computing power to the edge computing unit through the network architecture of the cloud side, and then completes a certain degree of accurate and lossless video data selection in the sensing front end through the completion of massive unstructured video data. This network architecture scheme can not only effectively reduce the transmission pressure of network bandwidth. It will also save system storage and computing resources, improve the real-time response speed and analysis accuracy of the system, reduce the system delay, and achieve efficient and timely response [7].

The security cloud edge end architecture scheme is shown in Fig. 1. The image and video data are collected through the camera, and then the target face frame in the photo is detected through local edge computing and features are extracted. The recognition results are fed back to the camera for output through feature matching with the cloud end.

2.2   Target Detection in Video Image Sequence of Wireless Visual Sensor Networks

The application scenario of the wireless visual sensing network identity feature recognition method proposed in this article is mainly video surveillance systems. Video surveillance systems have the characteristics of simple background and high real-time requirements. The original data of the dataset used in this article is also videos with relatively simple background, and the background image has not changed much, whether it is the experimental environment or the actual application environment of this article, The

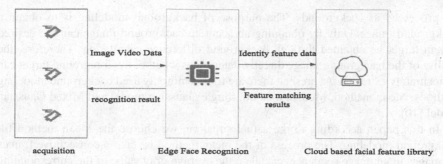

Fig. 1. Security Cloud Edge Architecture

background subtraction method is a good choice due to its performance advantages [8]. Therefore, this article chooses the background subtraction method as the basic algorithm for extracting human targets. For each image in the dataset, a rectangular box is used to mark the area where the human target is located and extract it.

The process of human object extraction is shown in Fig. 2, which can be divided into seven steps, namely image sequence acquisition, background modeling, background difference operation, binarization, open operation, connected domain analysis and target region extraction. Among them, open operation and connected domain analysis can be selected according to the actual situation.

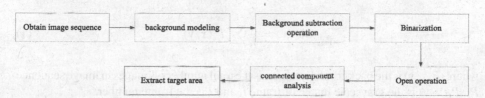

Fig. 2. Process of background difference method

(1) Obtain image sequence.

The data obtained from surveillance videos are all in video format, and the data set adopted in this paper is also in video format. The image sequence corresponding to each video can be obtained by extracting video frames from the wireless visual sensor network. For the background video provided in the data set, frame by frame extraction is adopted in this paper to ensure that higher quality background images can be obtained in the process of background modeling. For the video of wireless vision sensor network containing human objects, extraction is carried out according to the number of 5 frames per second, so that the image sequence can contain the main gestures in the walking process. In terms of quantity, it can also meet the training needs of convolutional neural network, and at the same time avoid excessive redundant data leading to increased computation and reduced training efficiency [9].

(2) Background modeling.

After obtaining a wireless visual sensing network video image sequence through video, the desired human target area is the foreground, while other areas of the image are processed as backgrounds. The purpose of background modeling is to obtain a background image. Only by obtaining an accurate background image can the desired human target be obtained through background difference calculation. Therefore, the quality of the background image directly determines whether the foreground target can be accurately extracted. At present, there are four commonly used background modeling methods: Mean method, Median method, Single Gaussian model and Mixed Gaussian model [10].

In this paper, according to the actual situation, we choose the Mean method for background modeling. The process of the mean method is: take n consecutive frames of images in an image sequence, calculate the average gray value of the corresponding pixel, and the average value is used as the final gray value of the pixel at the same position in the background image. The reason for choosing this method is that the quality of the background modeling averaging method can meet the requirements of the algorithm, the calculation speed is fast, and it is more in line with the real-time requirements of the video surveillance system.

Taking a background video from the CASIA (Chinese Academy of Sciences Institute of Automation) database Dataset B, Dataset SURF and Dataset CBD as an example, this article uses the mean method to calculate the background modeling process as follows:

After obtaining the image sequence of the background video frame by frame, grayscale each frame to obtain the image sequence P_i and i as the frame numbers. Calculate the average grayscale values of all corresponding points in the image sequence P_i, and establish the background image. The calculation formula is:

$$B(x, y) = \frac{1}{n} \sum_{i=1}^{n} P_i(x, y) \tag{1}$$

where, $B(x, y)$ is the background image, n is the total number of images in image sequence P_i, $P_i(x, y)$ is the grayscale image of frame i, and i is the frame number.

Through calculation, a background image can be obtained for the image sequence of each background video, that is, each person has a grayscale background image in each perspective, and the same background image is used for different dressing or walking posture under the same perspective.

(3) Background subtraction operation

For an image sequence containing human targets, first convert each image in the wireless visual sensing network video image sequence into a grayscale image to obtain image sequence G_i, and then perform a difference operation with the corresponding background image of the image sequence to obtain a new image F_i. The calculation formula is:

$$F_i(x, y) = |G_i(x, y) - B(x, y)| \tag{2}$$

where, F_i is the video image sequence of wireless visual sensor network obtained after background difference calculation, G_i is the video image sequence of grayscale wireless visual sensor network, and B is the background image.

As shown in Fig. 3, this figure is a new image obtained by background error calcula-tion. It can be seen that the perfect foreground target cannot be obtained only by simple background error calculation.

Fig. 3. Image obtained by background subtraction method

(4) binarization

Binarization is a method often used in the process of image processing. Through binarization, an image can be divided into two areas that are either black or white, and the part of interest is usually set as white. The method of setting a threshold for the whole image is called the global threshold method. The image can also be divided into multiple areas, and each area sets a threshold, which is called the local threshold method.

In this paper, the global threshold method is used to set a global threshold T. If the gray value of a pixel is greater than T, the color of the pixel is set to white. If the gray value of a pixel is less than T, the color of the pixel is set to black, and the binary image sequence R_i is obtained. The image more accurately selects the area where the foreground target is located and displays it as white. The calculation formula is:

$$R_i(x, y) = \begin{cases} 1 F_i(x, y) \geq T \\ 0 F_i(x, y) < T \end{cases} \tag{3}$$

After many experiments, for Dataset B, Dataset SURF and Dataset CBD of CASIA database, threshold T was set to 40, which had the most ideal effect. Figure 4 shows the image after binarization. It can be seen that some noises in the figure need further processing, and some details of the human body are also missing.

(5) Open operation

After binarization processing, the image still cannot fully meet the requirements, because there will be foreground empty points and background noise points, especially the background noise points have a great impact on the subsequent work of this paper, so it is necessary to carry out morphological processing on the binary image obtained before.

Fig. 4. Binary Diagram

Morphological processing of image is to improve the quality of binary image by logical operation between structural elements and binary image. The two most basic operations of morphological processing are corrosion and expansion. Corrosion can eliminate noise points smaller than structural elements, and expansion can fill the holes in the target, but both corrosion and expansion will obviously change the area of the target region.

The combination of open and closed operations through corrosion and expansion solves this problem. In this paper, the binary image is opened, that is, the erosion operation is carried out first, and then the expansion operation is carried out. The main purpose is to remove some small noises in the binary image.

First, the corrosion operation is carried out, and the set obtained by etching the binary image R_i through the structural element S is the set of the origin position of S when the structural element S is completely included in the binary image R_i. The corrosion calculation is as follows:

$$D_i = R_i \odot S \tag{4}$$

where, D_i is the image obtained after corrosion operation, R_i is the binary graph, and S is the structural element.

After the expansion operation, the set obtained by the expansion of the binary image D_i through the structural element S is the set of the origin position of S when the displacement of S' intersects with at least one non-zero element in the binary image D_i. The expansion calculation formula is:

$$E_i = D_i \oplus S \tag{5}$$

In the equation, E_i represents the image after expansion operation.

Figure 5 (a) shows the image after corrosion operation on Fig. 4, and Fig. 5 (b) shows the image after expansion operation.

(6) Connected domain analysis.

The open operation can remove the noise with relatively small area, but the noise with relatively large area can be removed by connected domain analysis. Connected domain refers to the area formed by the connection of points with adjacent positions and equal pixel values in the image, and connected domain analysis refers to marking

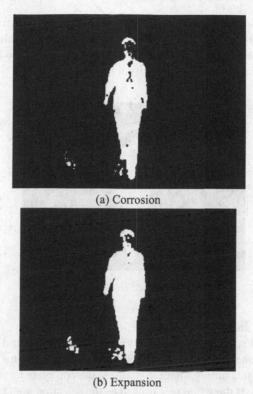

(a) Corrosion

(b) Expansion

Fig. 5. Corrosion and Expansion of Binary Graph

the white area in the binary image so that each single connected area has a unique mark, so that geometric parameters such as contour, centroid and external rectangle of these blocks can be obtained further. In this paper, pixel labeling method, which is widely used in connected domain analysis, is used.

Scan all pixels in the binary graph and assign a unique label to the set of pixels located in the same connected region. During the scanning process, there may be multiple small connected regions assigned different labels, but these small connected regions belong to a larger connected region. Write these small connected region labels into equivalent pairs and record their equality relationship.

Merges equal connected domains into one connected domain and assigns a new tag. The area of each connected domain is calculated, and the connected domain whose area is less than a certain threshold is regarded as noise for removal. Through many experiments, it is found that for the data set used in this paper, the effect of setting the threshold as 125 pixels is ideal.

The image obtained after connected domain processing is shown in Fig. 6. It can be seen that the noise with a relatively large area is also removed. Although some details are lost in the foreground area, it can meet the need of extracting the human body target in the video image sequence of wireless visual sensor network.

Fig. 6. Connected Domain Analysis

(7) Extract target area.

After background subtraction, binarization, open operation, and connected domain analysis, the background noise of the color image containing the target character is eliminated, resulting in a binary image containing the complete target, where the white area is the target area. If the white part is used as the selection area to extract the target, as shown in Fig. 7 (a), there will be some loss in the details, especially in the facial details, which will have a certain negative impact on the accuracy of the recognition results. Therefore, the method adopted in this article is to calculate the outer rectangle of the target character area through the outline of the white area, and use the internal area contained in the rectangle as the selection area of the target character, Through this selection area, a rectangular area can be captured from the original color image, which includes both the human target and some background images, as shown in Fig. 7 (b). The rectangular area is used as the final human target to be extracted.

2.3 Recognition Model Based on Convolutional Neural Network

Convolutional neural networks (CNN) have evolved from traditional neural networks, with the main difference being that the feature extractor of convolutional neural networks is mainly composed of convolutional feature extractors, while traditional neural networks are mainly composed of fully connected layers. The structure of CNN network used in this paper includes three main parts: convolution layer, activation layer and pooling layer. Among them, the convolutional layer is one of the core components in CNN, which extracts the local features of the input image through the convolution operation. The convolutional layer consists of multiple kernels, each of which multiplies element-wise with a local region of the input image and sums the results to obtain an element of the output feature map. The activation layer introduces nonlinear transformation to increase the expression ability of the network. Commonly used activation functions include ReLU

(a) Human targets

(b) Rectangular area

Fig. 7. Human target area

(Rectified Linear Unit), sigmoid, and tanh. The activation layer usually follows the convolutional layer and performs element-wise activation function computation on the output of the convolutional layer. The pooling layer reduces the size of the feature map by downsampling operation while retaining important feature information, and maps the pooling results to the output layer to realize the identification of identity features. In this paper, convolutional neural networks are used to construct Softmax classifier to classify human body targets in video image sequences of wireless visual sensor networks, so as to identify identity features.

2.3.1 Convolutional Layer

In the traditional fully connected network, each node will connect all nodes in the upper layer, which will lead to excessive parameters, difficulty in model training, and overfitting. In convolutional neural networks, the convolutional layer implements local connections, which can effectively reduce the number of model parameters. Each neuron in the convolution layer is only connected to the local Receptive field of the previous layer. The size of the local Receptive field depends on the size of the convolutional nucleus. In CNN convolution, two-dimensional convolution is defined as follows:

$$s(i,j) = (X \times W)(i,j) \tag{6}$$

It can be seen from the above equation that the convolution result of the human object image in the video of wireless visual sensor network is the result of multiplying the local region of the image and the elements of each position of the convolution kernel matrix, and then adding. It is assumed that the size of the feature mapping image input at layer l is $W \times H \times D$, where W and H are the width and height of the human object identity feature image in the two-dimensional wireless visual sensor network video, and D determines the depth of the feature image. Parameter P, the number of 0 elements filled around the image, is used to adjust the size of the output identity feature map. If the size of the two-dimensional convolution kernel is $W_c \times H_c$, the number of output channels set is k, and the step size is S, then the size of the feature map obtained after convolution is $W' \times H' \times D'$. We can control the size of the feature image after convolution by using the convolution kernel size, step size S, and zero padding P. For example, if the convolution kernel size is set to 3, P is set to 1, and step size is set to 1, the feature image remains the same size after passing through the convolution layer. If the size of the convolution kernel is set to 2, P is set to 0, and the step size is set to 2, the feature image will be reduced to 1/4 of the original to achieve the effect of Downsampling.

2.3.2 Activation Layer

Convolutional layers are essentially linear, but for sample data that needs to be learned, their distribution may not necessarily be linearly separable. In order to learn the nonlinear part, an activation layer is usually connected behind the convolutional layer. The activation function in the activation layer transforms the data nonlinearly, which is the key for neural networks to solve nonlinear problems. Several common activation function are sigmoid, tanh, Re LU.

The sigmod activation function has two main disadvantages: (1) It is easy to be saturated. When the input is very large or very small, the gradient is approximately 0, which will lead to the gradient dispersion in the back propagation. (2) The non zero mean output of sigmoid will affect the output of the back layer, thereby affecting the update of the gradient: if the inputs of the back layer neurons are all positive, the local gradient obtained is positive. During the backpropagation process, the parameters will always update in the positive direction, and vice versa, the parameters will always update in the negative direction, resulting in slower convergence speed.

tanh activation function compresses the output to the range of -1 ~ 1, and its output basically follows the mean value distribution of 0, but the problem of gradient saturation still exists in tanh function.

The Re LU activation function does not require exponential calculation, and its computational complexity is relatively low. It only inhibits negative input values, and its output is sparse. Re LU solves the problem of gradient saturation in sigmod function. In this paper, Re LU is selected as the activation function, and its expression is as follows:

$$f(x) = \max(x, 0) \tag{7}$$

2.3.3 Pooling Layer

In the deep convolutional neural network, a huge amount of parameters will be generated with the deepening of the network, and the hardware conditions limit the infinite increase of parameters in the convolutional neural network, which requires the control of the number of parameters in the network. In addition, the image is static, and the same feature may apply to different areas in the image, which indicates that there must be redundancy in the original parameters. The essence of the pooling layer is to aggregate statistics of features of different locations and compress the input wireless visual sensor network identity feature map, that is, to conduct downsampling of the feature map. Pooling layer can effectively reduce the parameters required by subsequent layers, reduce the possibility of overfitting, and make the convolutional neural network translation invariant, that is, when the pixels in the feature map have a small displacement in the neighborhood, the pooling layer can keep the output unchanged, which enhances the robustness of the network. The commonly used pooling layer downsampling methods include mean pooling and max pooling. Pooling of regional average values can preserve the characteristics of the overall data and highlight the background information. Pooling the maximum value of a region can better preserve the features on the texture. The pooling methods used in this article are regional maximum pooling and mean pooling.

2.3.4 Softmax Classifier

Softmax is a multinomial logic regression model. Logistic regression model belongs to log-linear model and is also a probabilistic model, which is generally used for binary classification problems. Multinomial logistic regression model is the extension of logistic regression and can be used to solve multi-classification problems. The calculation process of Softmax is as follows:

$$P(Y|x) = \frac{\exp(w_k \times x)}{\sum_{k=1}^{K-1} \exp(w_k \times x)} \tag{8}$$

In the equation, w is the desired parameter model, x is the input vector, and $P(Y|x)$ is the probability value of predicting the category of x as category Y.

Input human target identity features from wireless visual sensing network videos as samples into Softmax classifier to obtain classification results for identifying wireless visual sensing network identity features:

$$\theta_t = -\eta \times g_t \tag{9}$$

where, η is the global learning rate initially set, and g_t is the gradient.

3 Experimental Analysis

3.1 Preparation for Experiment

In order to verify the effectiveness of the wireless visual sensing network identity feature recognition method based on convolutional neural networks proposed in this article in practical applications, the accuracy and recognition time of the wireless visual sensing network identity feature recognition were selected as experimental indicators, and the methods of reference [4] and reference [5] were used as comparative methods for experimental testing.

This article also used GPU for acceleration during the experimental process. The GPU brand model is NVIDIA Tesla K40c, which has 2880 Cuda cores and 12G graphics memory. Its parallel computing and storage capabilities can well meet the needs of convolutional neural network training. Using M3001 robot module camera, capturing image size 1920 × 1080, which can set parameters such as automatic exposure, automatic white balance, color correction, brightness, contrast, saturation, sharpness, etc. It supports a variety of protocols such as common TCP/IP, ICMP, HTTP, FTP, DHCP, DNS, DDNS, RTP, RTSP, etc. It can access the network through RJ45 10M/100M adaptive Ethernet port, and transmit the photos taken to the local processing unit. The camera and processor are shown in Fig. 8.

Fig. 8. Camera and processor

The Hiss 3516DV300 neural network processing chip used in the terminal computing unit in this paper NNIE, short for Neural Network Inference Engine, is a hardware unit specializing in accelerated processing of neural networks, especially deep learning convolutional neural networks in So C of Hesis Media. It supports most existing public networks. For example, classification networks such as Alex Net, VGG16, Res Net18 and Res Net50, detection networks such as Faster R-CNN, YOLO and SSD, and scene segmentation networks such as Seg Net and FCN.

3.2 Data Set Construction and Training

(1) Building a dataset.

Obtaining multi view video data of all objects to be identified through the reasonable arrangement of monitoring equipment. From a practical application perspective, it is reasonable to have no more than three views. Then, the video data is converted into an image sequence and target detection is performed using background subtraction to extract the rectangular area where the human target is located as the dataset. If image data from three different perspectives is obtained, the dataset is divided into three subsets according to three different perspectives, and each subset is further divided into a training set, a validation set, a single perspective testing set, and a multi perspective testing set.

(2) Training model.

The multi-network identity model is trained and tested on the multi-view data set according to the training and testing process. Firstly, an appropriate convolutional neural network should be built as the subnet for training each perspective, and the training set and verification set of each subset should be input into the corresponding subnet for training, so as to obtain the model of each subnet. During the training process, each subnet should be adjusted independently, so that each subnet can obtain the best recognition result on its corresponding data set. The accuracy of each subnet is tested by the single view test set, and the weight of each subnet is calculated according to the weight calculation formula, and the weighted fusion identity model is obtained. The final identity model is tested through the multi-view test set to verify its comprehensive performance and further optimize the model.

(3) Identity recognition.

Input the multi view image of the object to be recognized into the corresponding subnet for recognition, obtain the recognition results of each subnet, and then calculate the final identity recognition result through weighting.

CASIA database, created by the Institute of Automation of the Chinese Academy of Sciences, consists of three data sets, of which Dataset A is a small-scale database with 20 people, each with three shooting angles (0°, 45°, 90°), a total of 240 image sequences, and the acquisition environment is outdoor. Select Dataset B, Dataset SURF and Dataset CBD of the CASIA database as the dataset for this article. Each subject has eleven perspectives, and the images from different perspectives are shown in Fig. 9.

From the different perspectives mentioned above, the methods of this article, reference [4], and reference [5] were used to compare and analyze the accuracy of identity feature recognition in wireless visual sensing networks. The comparison results are shown in Table 1.

According to Table 1, the accuracy of the method used in this paper for identity feature recognition in wireless visual sensing networks can reach up to 99.5%. The accuracy of the method used in reference [4] for identity feature recognition in wireless visual sensing networks can reach up to 84.1%. The accuracy of the method used in reference [4] for identity feature recognition in wireless visual sensing networks can reach up to 70.5%. The accuracy of the method used in this paper for identity feature recognition in wireless visual sensing networks is the highest, and the recognition effect is the best.

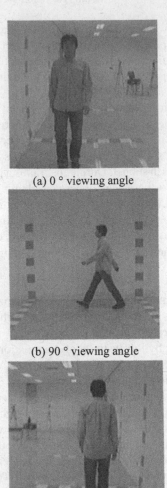

(a) 0 ° viewing angle

(b) 90 ° viewing angle

(c) 180 ° viewing angle

Fig. 9. Images from Different Perspectives

Using the methods of this article, reference [4], and reference [5], a comparative analysis was conducted on the time required for identity feature recognition in wireless visual sensing networks. The comparison results are shown in Table 2.

As can be seen from Table 2, the time used for the identification of wireless visual sensor network identity features by the method in this paper is within 6.2s, the time used for the identification of wireless visual sensor network identity features by the method in reference [4] is within 16.5s, and the time used for the identification of wireless visual sensor network identity features by the method in reference [4] is within 26.4s. The method presented in this paper has the shortest time and the highest recognition efficiency for wireless visual sensor network identity feature recognition.

Table 1. Comparison results of identification accuracy of wireless visual sensor networks /%

Number of experiments/times	Textual method	Method of reference [4]	Method of reference [5]
10	94.2	80.1	64.5
20	94.9	80.6	65.2
30	95.6	81.2	66.2
40	96.8	81.6	67.4
50	97.5	82.3	68.1
60	98.2	83.5	69.2
70	99.5	84.1	70.5

Table 2. Comparison results of identity feature recognition time in wireless visual sensor network /s

Number of experiments/times	Textual method	Method of reference [4]	Method of reference [5]
10	5.2	15.2	22.2
20	5.3	15.6	22.6
30	5.4	15.7	23.4
40	5.5	15.9	23.5
50	5.5	15.9	24.9
60	5.8	16.2	25.8
70	6.2	16.5	26.4

4   Conclusion

In recent years, with the gradual development and widespread application of wireless visual sensing networks, people's requirements for information security have become increasingly high. Identity verification, as one of the important means to ensure information security, can advantageously ensure that system users have corresponding application rights. Therefore, studying adaptive recognition algorithms for identity features is of great significance and has become a key research topic for relevant scholars, receiving increasingly widespread attention. As a key issue in the development of network security, identity recognition technology has received increasing attention from scholars. Commonly used identity recognition technologies mainly include face recognition, iris recognition, fingerprint recognition, and related algorithm research has achieved certain results.However, in the research of visual optimization identification, due to the influence of posture, light, expression and other factors, it is impossible to accurately identify the identity in the wireless visual sensor network under the uncontrollable environment. In traditional visual identity recognition, uncontrollable factors need to be transformed into controllable and stable characteristic factors in an uncontrollable environment with relatively complex node distribution before identity recognition. The conversion process leads to long recognition time and low efficiency. In this paper, an identity feature recognition method based on convolutional neural networks is proposed for wireless visual sensor networks, and the experimental verification shows that the proposed method has good identification effect and high recognition efficiency. Future research should also focus on the use of identity feature recognition techniques for privacy protection and security. This paper explores how to fully consider the needs of user privacy and information security while ensuring high accuracy.

References

1. Zheng, Y.L., Burns, J.H., Wang, R.F., et al.: Identity recognition and the invasion of exotic plant. Flora Morphol. Distrib. Funct. Ecol. Plants **280**, 151828 (2021)
2. Yang, W.-H., Dai, D.-Q.: Two-dimensional maximum margin feature extraction for face recognition. IEEE Trans. Syst. Man Cybern. Part B (Cybern.) **39**(4), 1002–1012 (2009). https://doi.org/10.1109/TSMCB.2008.2010715
3. Dongbo, L.I., Huang, L.: Reweighted sparse principal component analysis algorithm and its application in face recognition. J. Comput. App. **40**(3), 717–722 (2020)
4. Hou, X., Li, R., Zhang, Y.: Personnel characteristics identification based on foot induced structural vibration. J. Vibr. Shock **41**(23), 241-248,292 (2002)
5. Yao, L., Guo, S., Yang, X.: Threshold identity authentication scheme based on biometrics. App. Res. Comput. **39**(4), 1224–1227 (2022)
6. Ding, W.: Network video surveillance equipment identification based on Web identity characteristics. J. Shenyang Univ. Technol. **42**(4), 427–431 (2020)
7. Zhao, D., Lu, Y., Liu, X., et al.: Design of emergency UAV network identity authentication protocol based on Beidou. MATEC Web Conf. **336**, 04004 (2021)
8. Yichao, Z., Ziwen, S.: Identity authentication for smart phones based on an optimized convolutional deep belief network. Laser Optoelect. Progr. **57**(8), 081009 (2020)
9. Tian, Z., Yan, B., Guo, Q., et al.: Feasibility of identity authentication for IoT based on blockchain. Proc. Comput. Sci. **174**, 328–332 (2020)
10. Liu, Y.N., Lv, S.Z., Xie, M., et al.: Dynamic anonymous identity authentication (DAIA) scheme for VANET. Int. J. Commun. Syst. **32**(5), e3892.1-e3892.13 (2019)

Feature Recognition of Rural Household Domestic Waste Based on ZigBee Wireless Sensor Network

Meitong Zhao and Xiaoying Lv[✉]

Dalian University of Science and Technology, Dalian 116052, China
lvxy1986@126.com

Abstract. The current feature recognition data set of rural household garbage is generally set as one-way, and the recognition range is greatly limited, resulting in an increase in the average difference of feature recognition. Therefore, the design and verification analysis of the feature recognition method of rural household garbage based on ZigBee wireless sensor network is proposed. According to the actual recognition requirements and changes in standards, the initial multi-scale fuzzy features are extracted first, and the form of multiple targets is used to expand the actual recognition range. Multi target cross recognition data sets are set up to build ZigBee wireless sensor network garbage feature recognition model, and anchor box clustering processing is used to achieve feature recognition. The final test results show that the average difference of feature recognition obtained from the identification and measurement of rural domestic waste characteristics at the selected five test points is well controlled below 1.5, indicating that this recognition form has strong controllability and pertinence, large recognition range and controllable error, and has practical application value.

Keywords: ZigBee Wireless Sensor · Domestic Garbage of Rural Households · Household Waste Feature Recognition

1 Introduction

With the rapid development of·economy in rural areas and the improvement of people's living standards, the quantity and types of household garbage in rural areas are increasing. The problem of domestic waste management in rural areas is increasingly prominent, which poses a serious threat to the environment and people's health [1]. Therefore, it is the key to solve this problem to identify and classify the characteristics of rural household garbage. Due to the large population base and unbalanced regional development in China, most of the garbage in some developed areas will be recycled harmlessly. In underdeveloped areas, such as remote villages or towns, the implementation effect of this work will be relatively poor, and the daily domestic garbage generated by rural households is often treated by landfill [2]. Although this method can achieve the expected garbage treatment goal, it lacks pertinence and stability, which will also have a great

L. Yun et al. (Eds.): ADHIP 2023, LNICST 550, pp. 199–213, 2024.
https://doi.org/10.1007/978-3-031-50552-2_13

impact on the quality and efficiency of garbage treatment, and even lead to a large number of land being occupied by garbage, which will lead to the pollution of soil and groundwater [3, 4]. In this background, the classification and feature identification of rural household garbage has gradually become an effective means to prevent garbage pollution and realize garbage recycling, which can not only realize the reuse of resources, but also bring economic benefits to society [5, 6]. Therefore, an accurate and efficient intelligent garbage classification model plays an extremely important role in improving the construction of rural ecological civilization.

Therefore, a feature recognition method of rural household garbage based on ZigBee wireless sensor network is proposed. As a new wireless communication technology, ZigBee wireless sensor network has the advantages of low power consumption, low cost, self-organization and self-repair, and is widely used in various fields. In the aspect of identifying the characteristics of rural household garbage, ZigBee wireless sensor network can realize real-time monitoring and data collection of garbage, which provides strong support for subsequent treatment and management.

The significance of this study is to improve the intelligent level of rural household garbage management and reduce the waste of human resources and environmental pollution by introducing ZigBee wireless sensor network technology. At the same time, through the identification and classification of garbage characteristics, it provides scientific basis for garbage reuse and resource utilization, and promotes sustainable rural construction. In different background environments, the monitoring points with the same recognition distance are set, the preset feature recognition range is expanded, and the recognition processing errors in the process are minimized, which provides reference and theoretical reference for the further development of post-related technologies and industries.

2   Design of ZigBee Wireless Sensor Network Feature Recognition Method for Rural Household Domestic Waste

Firstly, by extracting the initial multi-scale fuzzy features, the identification method enhances the feature expression ability of rural household garbage. Then a multi-target cross-recognition data set is designed, which contains different kinds and characteristics of garbage samples to improve the generalization ability of the recognition model. On this basis, a garbage feature recognition model based on ZigBee wireless sensor network is constructed, and the anchor box clustering method is adopted to optimize the recognition model to realize accurate recognition of garbage features. The specific structure is shown in Fig. 1:

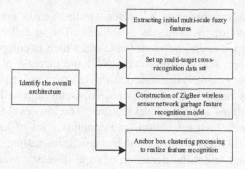

Fig. 1. Structural diagram of household garbage feature recognition in rural areas

2.1 Extraction of Initial Multi-scale Fuzzy Features

Before the feature recognition of rural household garbage, the initial multi-scale fuzzy features need to be extracted to provide reference for subsequent recognition processing [7, 8]. The multi-level recognition model is designed by combining ZigBee wireless sensor network technology and ResNet residual network [9, 10]. By adding a jump link layer between the input layer and the output layer, this method can effectively solve the problems such as gradient disappearance or gradient explosion of the convolutional neural network with the increase of the number of layers, effectively control a series of situations such as the rapid decline of network convergence and the reduction of the generalization ability of the model, and improve the recognition accuracy of garbage types and features. To a certain extent, it has improved the speed of training the ultra deep neural network, and the network performance has been greatly improved as the depth increases.

The principle of "identity mapping" is proposed so that the loss of training error will not increase when the number of layers is increased in ZigBee wireless sensor network. The original state is that when the input of ZigBee wireless sensor network changes, there will be no large error in the feature recognition result of rural domestic waste, and the change of residual function will also be effectively controlled. At this time, the basic identification unit value can be calculated, as shown in Formula 1 below:

$$B = O^2 \times \upsilon + \frac{O(1 - \sum_{t=1} \pi t + \upsilon)^2}{\omega} - p \tag{1}$$

In Formula 1: B Represents the basic identification unit value, O Indicates the coverage identification range, O Indicates the deviation of conversion identification, ω Represents the approximate value, π Represents the classification distance, t Represents the number of times of recognition, p Indicates the overlap range. According to the above measurement, complete the calculation of basic identification unit value. Next, integrate the actual garbage treatment requirements and changes in standards, adjust the identification structure and reference value, form a complete identification framework, and obtain fuzzy identification results.

Then, on this basis, combined with ZigBee wireless sensor network technology, a garbage feature recognition pyramid was established. The so-called feature pyramid is mainly a pyramid shaped neural network structure, which is composed of multi-level feature layers at different scale feature layers. With the increase of network depth, the scale of the feature layer gradually decreased, forming a pyramid structure, This structure significantly improves the detection performance of small targets. The feature pyramid contains low-level features and high-level features. The low-level feature map is close to the input image, with high resolution, and contains a lot of location information. The small receptive field leads to the lack of semantic information. The bottom features are used to identify and orient the detection of rural domestic waste features. On the contrary, high-level feature recognition contains a lot of semantic information, which is more important for large target detection. The low-level features and high-level features play a complementary role, so fusing the extracted information can improve the overall performance of the target detection algorithm. The specific recognition principle is shown in Fig. 2:

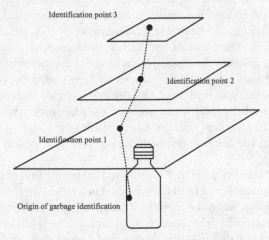

Fig. 2. Feature pyramid recognition results

According to Fig. 2, measure and analyze the recognition result of the feature pyramid. The forward calculation is carried out from the bottom up, and the corresponding execution process is designed. The high-level garbage recognition features are upsampled to improve the resolution, and the position information of each layer is correspondingly supplemented to the feature recognition structure through horizontal connection, so as to increase the receptive field, complete the recognition and collection of fuzzy features, and provide theoretical reference for the subsequent feature recognition work.

2.2 Set Multi Target Cross Identification Dataset

After the initial multi-scale fuzzy feature extraction is completed, then, combined with ZigBee wireless sensor network technology, the multi-target cross recognition dataset is set. You can take photos of rural domestic garbage first, and distribute all garbage pictures according to the number of garbage training set pictures as 85% of the total number of pictures, and the number of garbage test set pictures as 15% of the total number of pictures. Therefore, the garbage data set is divided into 155 garbage training sets and 1055 garbage test sets by code. When the data set is divided into training set and test set, it is randomly divided, so each classified garbage image in the garbage data set is randomly divided into the experimental training set and test set. In the garbage training set and garbage test set, the distribution of garbage images of these four categories is uneven. So in order to facilitate the analysis during the final test, this experiment will store all the garbage pictures of the same classification in the test set in the same garbage test set, and set the corresponding feature extraction indicators and parameters in combination with different domestic garbage, as shown in Table 1:

Table 1. Feature Extraction Indexes and Parameters of Different Domestic Garbage

Garbage name	Identification frequency/time	Time consumption/s	Recognition conversion ratio
Recyclable waste	12	1.02	3.2
kitchen waste	15	1.13	4.5
hazardous waste	16	1.05	4.1
Other Waste	10	1.06	4.6
Renewable waste	14	1.01	3.5

According to Table 1, complete the setting and adjustment of different domestic waste feature extraction indicators and parameters. The number of all classified garbage images in the garbage data set. The collected recyclable garbage images are summarized and classified into data sets. ZigBee wireless sensor network technology, Faster R-CNN and YOLOv3 methods should be used to design identification programs, and PASCAL VOC2007 data sets should be used. In order not to do repetitive work. So this time, we will use the analogy of PASCAL VOC2007 dataset to create our own garbage dataset. Then, store all collected garbage pictures in the JPEGImages directory, and convert all picture names into 6-digit numbers as shown in the above figure through script code. Because Faster R-CNN can only read JPG pictures, it converts PNG and other collected pictures with scripts.

The annotation directory stores the XML files that correspond to the images in the JPEGImages directory one by one. The XML files are generated after labeling with the labeling tool Labelling. The quality and quantity of garbage images in the dataset will have a great impact on the training results of the network. The number of pictures collected in the garbage dataset is not enough, which will lead to poor detection effect.

Therefore, in order to get better training weights and better test results, we need to expand the number of garbage datasets. The existing collected garbage images are expanded by image transformation to increase the sample size of garbage dataset and improve the robustness of the model. The scale or ambiguity of garbage features is processed by using filters. Use the oriented dimension tool Labelling to mark the specific location of the feature. Realize the design of basic data sets.

2.3 Construction of ZigBee Wireless Sensor Network Garbage Feature Recognition Model

After completing the setting of multi-target cross recognition dataset, the next step is to build ZigBee wireless sensor network garbage feature recognition model. In the absence of training data, garbage feature recognition models are extremely prone to over fitting, especially for compound garbage. Unable to systematically learn the characteristics of training data, which affects the overall cognition of the network model to the data, and ultimately leads to recognition errors or correlation recognition errors in the classification results. The root cause of the over fitting problem in the network is the lack of training data. Therefore, the data enhancement technology is used to expand the data of the domestic waste classification dataset DCD to solve the network over fitting problem from the source. In order to expand the training samples, the method of random clipping, color dithering, horizontal and vertical flipping and other spatial geometric transformations is used to supplement the data set.

Through the above methods, the common features in the training samples are enhanced, the stability of the neural network is improved, and the interference of noise information is weakened. Randomly cut the collected garbage images according to a certain size and length width ratio, and then scale the cut image to a fixed size image. Taking banana skin image as an example, the image can be obtained after random clipping. The resized image is a clipping image with different sizes and different aspect ratios. Then, a color jitter recognition program is added to the model. Color jitter is a data enhancement method for image color. It can reduce background interference by adjusting the brightness, saturation and ratio of the image to expand training samples. Take the banana skin image as an example, the image obtained after different adjustments of image brightness, contrast and saturation, and the result of special diagnosis recognition will show small changes after adjusting the three parameters at the same time.

Then the image is flipped horizontally and vertically. This part is mainly to transform the spatial position of image pixels, mirror the original image, and transform the direction of objects in the image to expand the training set. Take the banana skin image as an example, the image is obtained by vertical flipping after different adjustments. Based on this, the identification reference value is set according to the fixed characteristics of different types of garbage, as shown in Table 2:

Table 2. Setting of Reference Values for Fixed Features of Different Types of Garbage

Same type of garbage	Feature Enhancement Ratio	Training limit deviation value
Renewable waste	16.27	4.2
Non renewable waste	15.22	5.1
kitchen waste	14.21	5.1
Decomposable waste	11.03	3.1
Non decomposable waste	16.35	3.6
Recyclable waste	15.24	4.5
Non-recyclable waste	10.24	2.3
Other garbage	11.35	6.1

According to Table 2, set and analyze the benchmark values of fixed characteristics of different types of garbage. Then, based on this, the basic program of garbage feature recognition model was designed by integrating ZigBee wireless sensor network technology. It can be divided into four main contents as follows:

(1) Conv layers comprehensive recognition processing. This is a target detection method. In Faster R-CNN, the garbage image is used as input, imported into the model, and finally the feature map of the image is output to prepare for later use and identification analysis.

(2) ZigBee wireless sensor network. ZigBee wireless sensor network first judges whether the anchor frame is foreground or background through softmax. Then, the bounding box regression method is used to modify the anchor box to get more accurate candidate regions. Through ZigBee wireless sensor network technology, feature data and information are collected for future use.

(3) RoI pooling. The pooling layer of the region of interest processes the data such as the feature map and the candidate region, and obtains the feature map of the candidate region. Finally, the feature map is input into the full connection layer to calculate the category of target objects in the candidate area.

(4) Classifier + ZigBee wireless sensor network. The classifier uses the feature map of the candidate region obtained in the previous steps to determine the category of the candidate region. The ZigBee wireless sensor network technology is used to design a directional recognition mechanism, and finally the accurate position of the detection frame is obtained again using the frame regression. So far, the construction of ZigBee wireless sensor network garbage feature recognition model has been completed, and basic recognition results can be obtained.

2.4 Anchor Box Clustering for Feature Recognition

After completing the construction of ZigBee wireless sensor network garbage feature recognition model, the next step is to integrate ZigBee wireless sensor network technology and use anchor box clustering processing method to conduct feature recognition processing. ZigBee wireless sensor network borrows the anchor mechanism used in Faster R-CNN and SSD, and introduces an anchor box for detection. The settings of the anchor frame are different from those of Faster R-CNN and SSD. In Faster R-CNN and SSD, their settings for garbage feature recognition are manually set, while in Zig-Bee wireless sensor network, the settings of the anchor frame are k-means clustering operations on the target frame in the image.

ZigBee wireless sensor network is relatively limited. In the process of feature recognition, there are only five recognition anchor frames, so the number of k-means clustering classes k is 5, and finally the length and width of the five anchor frames are obtained. In ZigBee wireless sensor network, the number of anchor frames is increased. The number of k-means clusters k is taken as 10, and the length and width of 10 anchor frames are finally obtained. There are only 6 anchor frames in yolov3 tiny, so the number of k-means clusters k is 6.The center of each class after k-means convergence is the anchor frame. The principle of clustering feature recognition is to use pre labeled garbage frames, and automatically count the output of all labeled frames as the length and width of the anchor frame.

Compared with the use of Euclidean distance, the IOU used will not bring errors due to the different size of the bounding box. So in ZigBee wireless sensor network, IOU is used to calculate distance. And design the corresponding constraint conditions for anchor box clustering household garbage feature recognition, as shown below:

(1) First, 10 real feature tagging frames are randomly selected as the center of initial clustering recognition.
(2) Then, based on ZigBee wireless sensor network technology, for each other feature tag box, calculate and first select the directional recognition distance between the 10 cluster centers, and divide the tag box according to the distance.
(3) Then the average length and width of all feature callouts in each class are calculated respectively, and the combination of length and width is taken as the new class recognition center.
(4) Then, combined with ZigBee wireless sensor network technology, calculate the distance between all other label boxes and the 10 cluster centers obtained in the previous step. Divide the dimension box according to the conversion distance of the feature, and repeat the third step.
(5) Finally, stop the repeated operation until the change of the cluster center in step 4 is very small or the repeated operation is repeated for a certain number of times. At this time, the length and width of the 10 cluster centers are the size of the obtained prior box, which is the final recognition result of the garbage features. The specific structure is shown in Fig. 3:

According to Fig. 3, complete the processing and research of anchor box clustering processing feature recognition. The expected target of feature recognition can be achieved by using the designed model for detail processing and secondary recognition.

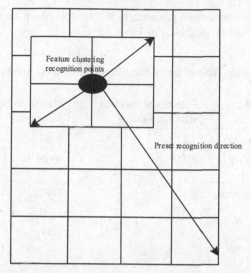

Fig. 3. Anchor box clustering processing feature recognition diagram

3 Method Test

This time is mainly to analyze and verify the practical application effect of the feature recognition method of rural household household domestic waste based on ZigBee wireless sensor network. Considering the authenticity and reliability of the final test results, the analysis is carried out by comparison, and H village is selected as the main target of the test, Use professional equipment and devices to collect basic test data and summarize and integrate information. According to the actual identification processing conditions and demand changes, compare and study the final test results. Next, combine ZigBee wireless sensor network technology to build the initial test environment.

3.1 Test Preparation

This time, ZigBee wireless sensor network technology is used to set up and build a multi-directional association with the test environment for the feature recognition method of household household garbage in H village. Combining the actual feature processing needs, the standard for lightweight garbage feature recognition and classification is built, the application mobile terminal for recognition is set, and the target detection algorithm is used to achieve the classification of multiple domestic garbage targets, It can improve classification speed. The location information of garbage needs to be obtained when using machines to grab garbage. The target detection algorithm can realize the classification and location of garbage. Finally, ZigBee wireless sensor network technology is used to complete the feature extraction of domestic garbage, and BiFPN is used to enhance the weighted bidirectional feature fusion of the extracted feature information to make full use of the underlying feature information and high-level semantic information, so as to further ensure the authenticity and reliability of the final recognition results. First, the initial identification indicators and parameters are set according to the actual domestic waste situation in Village H, as shown in Table 3:

Table 3. Setting Table of Initial Identification Indicators and Parameters

Initial identification indicators	Directional parameter standard value	Standard values of measured parameters
Convolutional stacking ratio	3.2	4.1
Single channel recognition rate/%	89.55	90.24
Gradient dispersion difference	6.3	7.4
Weight value	2.5	3
Localized directional recognition with poor range transition	0.62	0.44
Offset identification points/piece	12	18
Offset parameters	3.1	3.2
Identify center values	6.15	6.44

According to Table 3, complete the setting and adjustment analysis of initial identification indicators and parameters. Next, combined with ZigBee wireless sensor network technology, build a targeted domestic waste identification program, establish a stable identification matrix, and calculate the weight value at this time, as shown in Formula 2 below:

$$G = \left(1 - \chi^2\right) \times \frac{d\chi}{\beta + d - \lambda} + d^2 \tag{2}$$

Equation 2: G Represents the weight value, χ Represents the conversion ratio, β Represents an identifiable unit, d Represents the value of the search unit, λ Indicates the stacking range. Based on the above determination, the weight value can be calculated. According to the set feature recognition standards and conditions of domestic waste, a more stable feature recognition structure is constructed.

Then, on this basis, combined with ZigBee wireless sensor network technology, a multi-level, multi-target feature recognition structure was constructed. Build 25 layers 3×6 The convolution stack unit is localized and divided into five stages. The first layer of each stage recognizes the down sampling with a step of 2.5, and the activation function is ReLU. The main feature of RepVGG is that it decouples model training and reasoning, and adopts different network feature recognition architectures for different requirements in the training and reasoning stages of the ZigBee wireless sensor network applied. The most important thing in the training phase is accuracy, so ZigBee wireless sensor network technology adopts a multi branch structure, which consists of 3×6, 1×1 The multi branch structure is composed of a recognition standard combining convolution branch and residual branch of Identity. Different recognition differences are obtained by applying different wireless sensor units, and the information obtained from different recognition differences is fused to enhance feature extraction, so as to improve the performance of the feature recognition model of domestic waste.Next, according to the processing standard of ZigBee wireless sensor network technology, the total range of rural domestic waste is selected comprehensively to design the waste feature recognition process, as shown in Fig. 4:

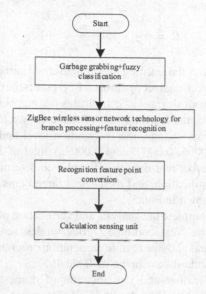

Fig. 4. Diagram of feature recognition process of rural domestic waste

According to Fig. 4, complete the design and application adjustment of the feature recognition process of rural domestic waste. Next, build a stable feature recognition training set through ZigBee wireless sensor network technology, and extend multiple gradient flow paths using residual units of multiple branches, which is equivalent to training multiple networks at the same time and fusion, similar to model integration. The ZigBee wireless sensor network identification model is obtained by equivalent transformation of the model in the training phase through structural re parameterization, and the single channel structure is used in combination with the sensor multi-directional and application network × 6 convolutional block composition, coupled with the sensing unit recognition matrix, greatly improves the speed of feature recognition for domestic waste, and facilitates model deployment and accelerated processing. In addition, in the ZigBee wireless sensor network model, feature information is obtained, and the ZigBee wireless sensor monocular is used as the backbone network. By changing the structure and parameters, the network model is converted from a multi branch structure to a single branch recognition structure, which improves the reasoning speed of the model. So far, the construction and correlation of the basic test environment have been completed. Next, the follow-up test and analysis will be made in combination with ZigBee wireless sensor network technology.

3.2 Test Process and Result Analysis

After completing the establishment of the basic test environment, next, based on ZigBee wireless sensor network technology, carry out measurement and verification research on the feature recognition method of domestic waste in H village. Select different learning rates to design a targeted feature world training matrix for domestic waste in the Faster R-CNN framework to form a controllable training model. Combined with ZigBee wireless sensor network technology, calculate the feature recognition learning rate at this time, as shown in Formula 3 below:

$$R = m + n \times \frac{m}{\sigma - \varsigma^2} - 1 \tag{3}$$

In Formula 3: R Indicates the learning rate of household garbage feature recognition, m Indicates the range of directional recognition, n Represents the unit identification value, σ Represents wireless sensor deviation, ς Indicates the number of studies. According to the above measurement, the learning rate of domestic waste feature recognition is calculated. According to its change, the recognition level and level are set, and the learning rate is 0.001, 0.0005, 0.00025 and 0.0001. A total of 20000 training sessions were conducted. Then use the real-time weight value to determine the fuzzy results of rural domestic waste identification, and then select three items as comparison objects. The specific results are shown in Fig. 5:

According to Fig. 5, complete the analysis and verification of the fuzzy results of the initial garbage identification. Next, combine ZigBee wireless sensor network technology to build a multi-dimensional garbage feature recognition program and establish multiple recognition levels and standards, as shown in Table 4:

Complete the identification and classification of domestic waste according to Table 4. Then, on this basis, combined with ZigBee wireless sensor network technology, the average difference of feature recognition is calculated, as shown in Formula 4 below:

$$D = \vartheta^2 - \left(\varepsilon + \sum_{y=1} \lambda y + c^2\right) - \varepsilon\vartheta \tag{4}$$

In Formula 4: D Represents the mean difference of feature recognition, ϑ Indicates the identifiable range, ε Represents the conversion fixed value, λ Represents the characteristic unit value, y Indicates the number of times that can be recognized, c Indicates the allowable limit difference. According to the above measurement, the five waste points in Xuanding Village H are taken as the test objectives to achieve the analysis of the test results, as shown in Table 5:

According to Table 5, the analysis of the test results is realized: the identification and measurement of the characteristics of rural domestic waste are carried out for the five selected test points, and the average difference of the final feature recognition is well controlled below 1.5, it shows that this identification method has good controllability and pertinence. At the same time, the identification method can cover a wide range of garbage types and characteristics, and the identification range is large. The error is also effectively controlled, which further proves the accuracy and stability of this method.

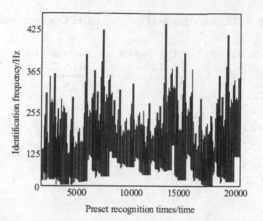

(a) Recognition results of rural household waste at a
learning rate of 0.0005

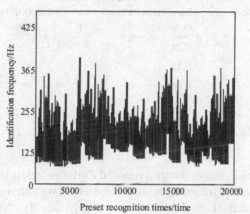

(b) Recognition results of rural household
waste at a learning rate of 0.00025

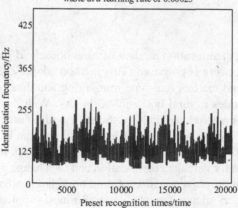

(c) Recognition results of rural household waste at a
learning rate of 0.0001

Fig. 5. Diagram of initial garbage identification fuzzy result analysis

Table 4. Identification and Classification of Domestic Waste

Identify categories	Learning rate 0.001	Learning rate 0.0005	Learning rate 0.00025	Learning rate 0.0001
Recyclable waste	325.6	9.3	14.5	16.3
hazardous waste	336.5	10.6	16.3	15.7
kitchen waste	342.1	106	18.5	10.8
Other Waste	352.1	11.3	19.5	17.5
Unrecognized object	359.5	12.6	21.3	19.4

Table 5. Comparison and Analysis of Test Results

Testing rural garbage sites	Identification time consumption/s	Mean difference in feature recognition
Garbage Point 1	0.24	1.3
Garbage Point 2	0.21	1.2
Garbage Point 3	0.26	1.4
Garbage Point 4	0.25	1.2
Garbage Point 5	0.22	1.3

The above research indicates that the proposed ZigBee wireless sensor network based feature recognition method for rural household household waste has practical application value. Through this method, rural areas can accurately identify and classify household waste, providing scientific basis for waste management and resource utilization.

4 Conclusion

In this paper, a feature identification method of rural household garbage based on Zig-Bee wireless sensor network is proposed. This method adopts ZigBee wireless sensor network technology, and realizes real-time monitoring and data collection of garbage by arranging sensor nodes in rural households' homes. With the help and support of ZigBee wireless sensor network technology, the structural accuracy that can be used to automatically identify the characteristics of household garbage in rural areas has been significantly improved. At the same time, the next research work needs to be combined with the actual demand for garbage classification and the change of standards. It is also necessary to strengthen the popularization of rural household garbage classification and environmental protection science knowledge at all times, supplemented by incentives and incentives, promote rural garbage classification, improve the rural environment, improve the humanistic quality of rural residents, and solve the problem of funds for construction and maintenance in many ways. Only in this way can the classification and treatment of rural household garbage be sustainable. In the future, we can further study and popularize the resource utilization technology of rural household garbage, and turn the garbage into renewable energy or valuable substances.

Acknowledgement. 1. Design of intelligent garbage sorting system based on AI machine vision-Innovation and Entrepreneurship Training Program for college students in 2023

2. Research on the Prediction method and system of base station network traffic based on Smart City-2021 Basic Scientific Research Project of Education Department of Liaoning Province (Youth Project)

References

1. Mudiyanselage, N.A., Herat, S.: Management of household hazardous waste: a review on global scenario. Int. J. Environ. Waste Manage. **2**, 29 (2022)
2. Sumbodo, B.T., Sardi, R.S., et al.: Analysis of the quadrant strategy for household solid waste management (case study: BUMDes amarta, pandowoharjo village sleman yogyakarta). In: IOP Conference Series: Earth and Environmental Science, vol. 739, no. 1, p. 012022 (8pp) (2021)
3. Liu, J., Ma, S., Shen, W., et al.: Image feature recognition and gas permeability prediction of Gaomiaozi bentonite based on digital images and machine learning. Adv. Geo-Energy Res. **6**(4), 314–323 (2022)
4. Zaharudin, Z.A., Brint, A., Genovese, A., et al.: A spatial interaction model for the representation of user access to household waste recycling centres. Resour. Conserv. Recycl. **168**, 105438 (2021)
5. Kurniawan, S., Novarini, Y.E., et al.: Combination of ozone-zeolite filter to Reduce COD and ammonia content in household waste. J. Phys. Conf. Ser. **1845**(1), 012071 (2021)
6. Rocha, P.F., Bezerra, A.L., Neto, J.M.: Household solid waste: an assessment of the impacts of covid-19 on preventive practices in Brazilian public management. J. Solid Waste Technol. Manage. **1**, 48 (2022)
7. Albira, I., Hamzah, T.A., Yusoff, S.M., et al.: Household solid waste generation: a case study of misrata city in Post-War Libya. Arab World Geographer **22**(1–2), 32–46 (2021)
8. Jung, S., Lee, M., Lee, S., et al.: A study on the trend of domestic waste generation and the recognition of recycling priorities in Korea. Sustainability **13**(4), 1732 (2021)
9. Zhao, S., Luo, J., Wei, S.: A hybrid eye movement feature recognition of classroom students based on machine learning. J. Intell. Fuzzy Syst. Appl. Eng. Technol. **2**, 40 (2021)
10. Liu, S., Zhang, D., Wu, Y., et al.: Wireless sensor network positioning algorithm based on RSSI model. Comput. Simul. **001**, 039 (2022)

A Method of Recognizing Specific Movements in Children's Dance Teaching Video Based on Edge Features

Chunhui Liu[1(✉)] and Chao Long[2]

[1] School of Economics and Management, Hunan Software Vocational and Technical University, Xiangtan 411000, China
liuchunhui869@163.com
[2] College of Art, Hebei University of Economics and Business, Shijiazhuang 050061, China

Abstract. Dance videos have issues with self occlusion and high complexity of actions, which affect the effectiveness of action recognition. In order to improve the accuracy of action recognition, a video action recognition method for children's dance teaching based on edge features is proposed. Image preprocessing for children's dance teaching videos, including grayscale and enhancement of video images. The background subtraction method is used to detect moving objects in video images, and the Canny operator is used to detect the edges of moving objects, enhancing the continuity of the edges. After obtaining an image that only includes the edges of the object, further extract the contour features of the object. A recognition method based on Adaboost BP neural network has been constructed. Using the BP neural network as a weak classifier, the Adaboost algorithm is combined with the outputs of multiple BP neural networks to construct a strong classifier, avoiding falling into local optima. Using edge features as input to achieve specific action recognition for children's dance teaching videos. The experimental results show that the recognition method based on edge features has a high average recognition accuracy of 94.715%.

Keywords: Edge Feature · Children's Dance Teaching Video · Pretreatment · Adaboost-BP Neural Network · Specific Action Recognition Method

1 Introduction

Motion capture recognition is very challenging in computer research, mainly using classification recognition and image processing technology to analyze video data to achieve human motion recognition. This direction has high research value and has attracted a large number of scholars and researchers from scientific research institutions. Motion recognition technology is applicable to a variety of video scenes, and has been widely used in video retrieval, intelligent human-computer interaction, virtual reality, intelligence, motion aided analysis and other fields. However, at present, the application of this technology in dance video is still less, and due to the problems of self occlusion and high

L. Yun et al. (Eds.): ADHIP 2023, LNICST 550, pp. 214–229, 2024.
https://doi.org/10.1007/978-3-031-50552-2_14

complexity of actions in dance video, the research in this area needs to be further carried out. The successful application of motion recognition technology in other fields also provides sufficient basis for the research in this field [1]. For the current large number of dance video data, professionals often need to spend a lot of time to analyze the data by watching and listening, which requires a lot of manpower and material resources and is extremely inefficient. The application of motion recognition technology in the analysis of these materials and the realization of dance motion recognition can not only reduce the work intensity of data analysts, facilitate the retrieval of video data, but also improve the efficiency of the automatic choreography system. It is also of great significance for mining and protecting cultural heritage in the art field and dance teaching [2]. In addition, the research in this area also has some reference and guiding significance for video action recognition in different environments, and can enrich the research in the direction of action recognition technology.

Human motion recognition has always been widely concerned. With the continuous progress of feature extraction methods and classification algorithms, motion recognition methods also continue to develop. Most of the early action recognition methods based on traditional computer vision use manual features extracted from action sequences, and then use multi-layer perceptron or support vector machine to classify based on these features. For example, based on the recognition method of improved dense trajectories, this method extracts features through the trajectories of sampling points, and then classifies the encoded features using SVM. This method has achieved good results in action recognition, and is one of the best algorithms in the field of traditional machine learning action recognition. However, traditional machine learning methods rely too much on high-performance manual design features, which requires a large number of experiments and prior information and is inefficient. With the increasing size of data sets and the increasing computing power of computing devices, deep learning has made great progress in the field of motion recognition. For example, the identification method based on dual stream network. The dual stream network is divided into two subnetworks, spatial stream network and temporal stream network. The spatial stream network extracts the spatial information of the sequence based on RGB image frames, and the temporal stream network obtains the temporal information of the sequence based on the optical flow extracted from adjacent image frames, and classifies them respectively. Finally, the average of the softmax scores of the two networks is taken as the classification result. The dual stream network is simple and effective, but it also has disadvantages correspondingly. In the time flow network, optical flow is the vector of motion. It is obtained by gradient calculation of two adjacent frames. In order to represent the motion information, it is input as multi frame optical flow. However, there are limitations. The number of frames is too small to describe the motion information well, resulting in poor recognition effect. If the number of frames is too large, the calculation time will increase, the efficiency will decrease, and the performance will not necessarily improve, Therefore, the effect of the two stream network on the long-term action modeling is poor. Another type of action recognition method is based on bone joint point data. Bone data is not easily affected by the above factors. Compared with RGB and depth data, the amount of bone data used to represent human action sequences is relatively small. Therefore, research on this type of method has gradually increased in recent years. For example, the method based on RNN converts bone data into feature vectors, and then models such as LSTM are used for modeling. Finally, the whole body bone features are sent into the classifier to get the final recognition results.

The main task of this paper is to study the methods of human motion recognition, and apply them to the recognition of specific dance movements, and propose a method of children's dance teaching video specific motion recognition based on edge features. The Canny operator is used to detect the edges of moving objects and enhance the continuity of the edges, thereby obtaining an image that only includes the edges of the object. This edge detection method can effectively extract the boundary information of objects. After obtaining the edge image of the object, a recognition method based on Adaboost BP neural network is used to further extract the contour features of the object. This method combines Adaboost algorithm and BP neural network, which can effectively improve the accuracy and robustness of the classifier. Innovatively using BP neural networks as weak classifiers and combining the Adaboost algorithm with the outputs of multiple BP neural networks to construct a strong classifier. This combination method can avoid falling into local optimum and improve the accuracy and stability of Object detection.

2   Research on Recognition of Specific Actions in Children's Dance Teaching Videos

The research on the combination of video motion recognition technology and dance art has just started in China. Through the application of human motion recognition technology to dance videos, dance movement posture can be effectively recognized. By comparing the video action with the standard action, we can evaluate the dancers' dance posture and give suggestions for modification, which is an advanced auxiliary training method.Under this background, this paper proposes a method based on edge features to recognize specific actions in children's dance teaching videos.

2.1   Pre Processing of Video Images for Children's Dance Teaching

(1) Video image graying

The grayscale transformation of an image refers to the method of changing the grayscale value of each pixel in the source image point by point according to the transformation function to achieve a certain target condition [3]. It can be expressed as

$$y(i,j) = I[f(i,j)] \tag{1}$$

where, $f(i,j)$ Is the grayscale function of the original image, $I[]$ Is a transformation function; $y(i,j)$ Is an output image function. The piecewise linear transformation can enhance the contrast of the image, highlight the areas of interest in the image, and effectively solve the problem of poor quality of the collected image. It is one of the commonly used gray transformation methods. The mathematical expression of the three-stage linear transformation method used in this paper is:

$$y(i,j) = \begin{cases} \frac{\chi}{\alpha} \times f(i,j) & 0 \leq f(i,j) \leq \alpha \\ \frac{\delta-\chi}{\beta-\alpha} \times [f(i,j)-\alpha]+\chi & \alpha \leq f(i,j) \leq \beta \\ \frac{\psi-\delta}{\psi-\beta} \times [f(i,j)-\beta]+\delta & \beta \leq f(i,j) \leq \psi \end{cases} \quad (2)$$

where, ψ It is the maximum gray scale of the video image. By adjusting the position of the inflection point of the polyline and the slope of the segmented straight line, that is, the control parameters α, β, χ, δ The expansion or compression of any gray range can be realized by taking the value of

(2) Image enhancement processing

In order to facilitate the subsequent analysis and processing by the computer, the video image of children's dance teaching is converted into a simpler gray image, but the gray image after conversion often has the problem of poor contrast, so the enhancement of gray image is crucial. Dynamic histogram equalization is developed on the basis of histogram equalization. Its equalization idea is to construct a cumulative distribution function through the probability density function of the original histogram, then use the cumulative distribution function and mapping interval to calculate the intensity value of the output image, and finally remap the pixel value of the original image [4]. The difference between the two is that dynamic histogram equalization divides the overall histogram into multiple sub histograms, and assigns a new mapping interval to each sub histogram. For interval: $[H_0, H_{L-1}]$ The probability density function and cumulative distribution function of a sub histogram of are as follows:

Probability density function:

$$\eta(k) = \frac{n_k}{A} \quad (3)$$

Cumulative distribution function:

$$\mu(k) = \sum_{q=H_0}^{k} \eta(q) \quad (4)$$

where, k It refers to a certain gray level within the range, n_k finger k The frequency of gray level appearance, that is, the number of pixels, A It refers to the total number of pixels in the range.

The specific process is as follows:

1) Input the video image of children's dance teaching.
2) Obtain the histograms of R, G and B channels of children's dance teaching video images respectively, which are recorded as H_R, H_G, H_B.
3) Calculation H_R, H_G, H_B The distribution range of.
4) According to the distribution range, the improved segmentation method based on exposure value is used to calculate the segmentation points iteratively, and each histogram is divided into four sub histograms.

5) Use the set reconstruction parameters b Calculate clipping threshold for each sub histogram T_i. The calculation formula is as follows:

$$T_i = \frac{A_i}{a_i} + b\left(f(H_i) - \frac{A_i}{a_i}\right) \tag{5}$$

Where, T_i On behalf of the i The total number of pixels in the sub histogram, a_i Refers to the length of the interval, b Is a reconstruction parameter whose value range is $(0,1)$, $f(H_i)$ Is to get the first i A function of the peak value of the sub histogram. Therefore, the adjustment range of the clipping threshold is between the average value and the peak value [5].

6) Statistics of exceeding clipping threshold in each sub histogram T_i Total number of pixels C_i And calculate its pixel redistribution value D_i.

$$D_i = \frac{(1-b)C_i}{a_i} \tag{6}$$

Among them, C_i Refers to the i The total number of pixels trimmed from the sub histogram. Reconstruction parameters are also used here b, so that the clipping threshold is associated with the reallocated value, that is, the more pixels are clipped, the larger the reallocated value is.

7) Rebuild each sub histogram, i.e.

$$E_i = \begin{cases} T_i, n_{ki} > T_i - D_i \\ n_{ki} + D_i, \text{otherwise} \end{cases} \tag{7}$$

Where, E_i Refers to the reconstructed No i Sub histogram, n_{ki} Is the first i In the sub histogram k The number of pixels on each grayscale.

8) Construct the corresponding cumulative distribution function for each reconstructed sub histogram $\mu(k)$.

9) A new mapping interval is allocated according to the total pixel proportion of each sub histogram after reconstruction.

$$H_i = \frac{Lg(E_i)}{\sum\limits_{g(E_i)=1}^{4} g(E_i)} - 1 \tag{8}$$

where, H_i Is the first i The segmentation point of the sub histogram mapping interval, g_q It is the first time after reconstruction q Sub histogram E_i The total number of pixels for.

10) Equalize each sub histogram independently and remap each channel.

$$G(k) = H_0' + (H_{L-1}' - H_0')\mu(k) \tag{9}$$

where, $[H_0', H_{L-1}']$ It refers to the new mapping interval.

11) The three channels after remapping are fused to obtain enhanced results.

12) End.

2.2   Edge Feature Extraction

In the research of action recognition, the first step is usually feature extraction. Feature extraction refers to extracting the feature information used to describe the target action in the video from the action data set, which is an essential step for the research of action recognition. From this point of view, the extracted features play a vital role in the accuracy of action recognition results and the robustness of action recognition methods. In this paper, after fully considering the characteristics of dance movements, edge features are extracted from dance videos. Before edge feature extraction, the moving human body needs to be extracted. A video image generally contains two parts: moving area and still area. The purpose of moving object detection is to successfully separate the two parts and extract the moving area [6]. Background subtraction is a common method for moving object detection in video images. This method is a method based on background modeling. First, a stable background is established through parameter updating, and then the original video object is compared with the background to get the detection result, i.e. foreground target. The flow of this algorithm is shown in Fig. 1 below.

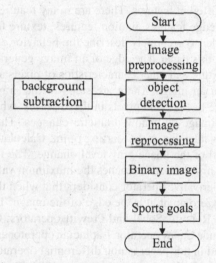

Fig. 1. Flow chart of background subtraction method

The above flow chart can be explained by formulas (10)~(11).

$$Q_t = |l_t - e_t| \tag{10}$$

$$\hat{Q}_t = \begin{cases} 1, Q_t > \varepsilon \\ 0, \text{otherwise} \end{cases} \tag{11}$$

In the above formula, l_t Represents the original video frame, e_t Is the background frame, Q_t by l_t and e_t The difference image obtained by subtraction. Equation (11) indicates Q_t The threshold value is set to ε, $\hat{Q}_t$ It is the result of binarization.

There is no doubt that the key of background subtraction based on background model is the establishment of background model. The common background modeling method is the average method. The principle of the average method is simple. It is to accumulate and sum multiple video image sequences containing moving objects in turn, and then get the background image [7] by averaging. The mathematical formula is shown in Eq. (12).

$$B(i,j) = \frac{\sum_{i=1}^{N} J(i,j)}{N} \tag{12}$$

where, N Represents the number of frames of the video image; $J(i,j)$ Represents the grayscale value of the image, $B(i,j)$ Is the desired background image, changing pixels (i,j) A stable and reliable background image can be obtained by using the value of.

In the above research, the moving human body has been successfully extracted, and the detection results obtained are all expressed in binary images. To further analyze the human behavior, it is necessary to describe a person's behavior as accurately as possible, which requires that some features that can fully represent the behavior be extracted first. Different images have different features. There are many features that can be extracted from an image, such as edge features, motion features, texture features, color features, light features, etc. In order to accurately describe the behavior of moving objects, edge features are extracted in this paper. The edge of an image generally refers to the places where the gray level, texture and other characteristics of pixels appear to be distributed in a jumping manner in an image. There are steps in the gray level changes of the image subject and image background pixels in these places, which are reflected in the function. The function image will show dramatic changes. Therefore, the traditional edge detection operators take this as the starting point. Calculate the first derivative or the second derivative [8] of the image gray level change. The first derivative operator considers that when the first derivative reaches the maximum value, it is the edge of the image, and the second derivative operator considers that when the second derivative of the function crosses the zero point, it is the edge of the image. The common first-order derivative operators are Roberts, Sobel and Prewitt operators; Common second-order derivative operators include Canny operator, Laplacian operator and Log operator. Canny operator to be introduced in this paper is non differential operator.

Canny gave three basic edge detection rules in 1986 as the basic idea of Canny operator, which is an optimized edge detection method up to now. The three basic principles are as follows:

① Signal to noise ratio rule: ensure the accuracy of the original image edge to prevent false edges;

② Positioning accuracy rule: the edge of the original image should be as close as possible to the tracked edge image;

③ Single edge response rule: the edge response should be unique to prevent multiple responses and resist virtual responses as much as possible.

The above three rules were first proposed by Canny, and Canny operator solved this problem completely in the form of mathematical expressions. The application of Canny operator in image edge detection has a very significant effect, which solves the problem of two-dimensional differential losing edge direction information, making Canny operator one of the most widely used algorithms in edge detection. The detailed operation process of Canny operator is as follows:

In the first step, Gaussian filter operator is used to smooth the noisy image. Because the edge and noise information in the image are mostly concentrated in the high-frequency part, the noise information is easy to be detected as a false edge in the image; To eliminate noise interference, the traditional Canny algorithm uses a two-dimensional Gaussian filter $\phi(i, j)$ The image edge is smoothed by convolution. Set the original image as $f(i, j)$, the smoothed image $O(i, j)$ It can be expressed as:

$$O(i, j) = \phi(i, j) * f(i, j) \tag{13}$$

In the formula "$*$" Represents the convolution operation, and the Gaussian filter function formula is as follows:

$$\phi(i, j) = \frac{\exp\left(-\frac{i^2+j^2}{2\gamma^2}\right)}{2\pi\gamma^2} \tag{14}$$

where, γ Is the standard deviation of the Gaussian function.

The second step is to obtain the gradient amplitude and gradient direction of all pixels in the image. After the smooth image is obtained through the Gaussian filter, the pixels of the image are in the horizontal direction x And vertical direction y Solve the partial derivative, and use the first order finite difference to calculate the gradient amplitude $Z(i, j)$ And gradient direction $\vartheta(i, j)$:

$$Z(i, j) = \sqrt{\left[\phi_x(i, j)\right]^2 + \left[\phi_y(i, j)\right]^2} \tag{15}$$

$$\vartheta(i, j) = \arctan\frac{\phi_x(i, j)}{\phi_y(i, j)} \tag{16}$$

The calculation formula for gradient amplitude and bearing transformation using rectangular coordinates is:

$$\phi_x(i, j) = \frac{f(i+1, j) - f(i, j) + f(i+1, j+1) - f(i, j+1)}{2} \tag{17}$$

$$\phi_y(i, j) = \frac{f(i, j+1) - f(i, j) + f(i+1, j+1) - f(i+1, j)}{2} \tag{18}$$

At this time, the edge strength of the image can be reflected, and the direction perpendicular to the edge can be reflected.

The third step is to perform non maximum suppression operation on the gradient amplitude of pixels in the image. Non maximum suppression is a kind of edge thinning technology. The role of non maximum suppression is to "thin" edges. After gradient calculation of the image, the edge extracted only based on gradient value is still fuzzy. For Criterion 3, there is and should be only one accurate response to the edge. Non maximum suppression can help suppress all gradient values other than the local maximum to 0. The algorithm for non maximum suppression of each pixel in the gradient image is:

(1) Compare the gradient intensity of the current pixel with two pixels along the positive and negative gradient directions;
(2) If the gradient intensity of the current pixel is the largest compared with the other two pixels, the pixel will remain as an edge point, otherwise the pixel will be suppressed.

In the fourth step, the double threshold algorithm is used to detect edges and connecting edges. Select high threshold first R_{max} And low threshold R_{min}, then start scanning the map $f(i, j)$ Each pixel of the candidate edge points marked in the candidate edge image (i, j) Detect if the pixel (i, j) Amplitude of $Z(i, j)$ greater than R_{max}, then this point is the edge point; If pixels (i, j) Amplitude of $Z(i, j)$ lower than R_{min}, then the point is not an edge point; gradient magnitude $Z(i, j)$ The points between high and low thresholds are regarded as suspected edge points, which need to be further judged according to edge connectivity. If there are edge points in the adjacent pixel points of the point, the point is regarded as an edge point for connection; Otherwise, the point is a non edge pixel and is discarded [9].

After the image containing only the target edge is obtained, the contour features of the object can be further extracted. The contour features that can be extracted are as follows:

1) Contour area: traverse all pixels of the image. If the pixel is in the contour, the value is increased by 1, and the final value is the contour area;
2) Contour Perimeter: calculate the spacing between adjacent points of the contour and accumulate it. The accumulated value is the contour perimeter;
3) Hu moment similarity: the second and third order normalized central moments of the object contour are linearly combined to obtain the Hu moment invariant of the contour, and the Hu moment similarity is obtained by comparing the Hu moment invariant of the target and the model;

2.3 Specific Action Identification

After the timely detection and effective feature extraction of moving objects are completed, the next problem is how to judge and classify the processed video image according to these effective features. The commonly said "identifying something" actually includes two points of recognition and distinction, that is, first recognize the object, that is, carefully observe one, two or more of its features, and then use these features to distinguish it from other objects in a pile of objects according to certain discrimination rules [10].

The full English name of BP neural network is back propagation neural network, that is, back propagation neural network. On the premise that the number of neurons in the hidden layer can be adjusted at will, it can approach any nonlinear mapping, and has a certain fault tolerance ability, so it is suitable as a classifier. Although BP neural network has many advantages [11], it is difficult to establish a better neural network system, which is caused by its own shortcomings. It mainly includes:

1) It is difficult to determine the number of iterations of the algorithm, and may oscillate around the local minimum value, which will lead to slow convergence of the algorithm. This will lead to too many iterations and low learning efficiency.
2) The gradient descent method is used to minimize the error function. The error function may contain multiple local poles [12]

3) The number of hidden layer nodes is difficult to determine. Must be selected by experiment or experience.
4) The newly added training samples will affect the original training samples, that is, the weight distribution of the network is modified.

Aiming at the problems of BP neural network, a massive image recognition method based on Adaboost-BP neural network is proposed. BP neural network is used as a weak classifier [13], and Adaboost algorithm is used to combine the output of multiple BP neural networks to build a strong classifier, as shown in Fig. 2 below.

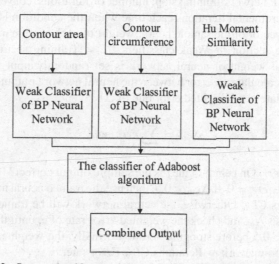

Fig. 2. Strong classifier model of the Adaboost BP neural network

AdaBoost (abbreviation of adaptive boosting) is a framework algorithm. AdaBoost learning algorithm can cascade a series of weak classifiers to form a strong classifier. The types of multiple weak classifiers used by AdaBoost classifier are consistent. Different classifiers are obtained through serial training, and later classifiers will be obtained through training according to the wrong data of the previous classifier. The output of AdaBoost classifier is the result of weighted summation of multiple weak classifiers [14]. The weight of each classifier is determined by its classification success rate. The higher the classification success rate, the greater the corresponding weight. Advantages of AdaBoost algorithm: low generalization error rate, easy implementation, and can be applied to most classifiers without parameter adjustment.

The construction of Adaboost-BP classifier needs two stages: training stage and recognition stage. The training process is as follows:

Input: training sample set V, the number of samples is n, one for each sample m Dimensional eigenvector U Representation, vector U The element in represents an edge feature of the sample and the number of categories is λ.

Step 1: Because the range of each feature of children's dance teaching video image is different, there is no comparability between features. In order to eliminate this difference, all characteristic values can be converted to the range of [0,1]. Therefore, before inputting the image feature vector of children's dance teaching video, it is necessary to normalize the image features:

$$\hat{v}_{ij} = \frac{v_{ij} - \min v_i}{\max v_i - \min v_i} \tag{19}$$

Among them, v_{ij}, $\hat{v}_{ij}$ It is the first time before and after processing j No. of samples i Features. max v_i, min v_i It is the No i The maximum and minimum values of the features.

Step 2: Set parameters of BP neural network weak classifier.

Step 3: Set the number of Adaboost iterations and training times ϖ The initial value is 0; Parameters of BP network training stop: number of iterations, convergence accuracy; The number of incremental iterations each time when the condition is not met N_u. Let the initial weight vector of the sample be w_0, weight of each sample $w_{i,0} = \frac{1}{n}$.

Step 4; Create section ϖ Weak classifiers CL_ϖ. Training section ϖ BP neural network, the initial weight of neural network is set randomly.input $\hat{v}_{ij}$ To the neural network, and then calculate the error between the neural network output and the expected output err_j, Calculate err_j Weighted error rate of Δe.

$$\Delta e_\varpi = \sum_{j=1}^{n} err_j \cdot w_{i\varpi} \tag{20}$$

Among them, err_j On behalf of the j Is the sample output correct? If it is wrong, then $err_j = 1$, conversely $err_j = 0$. If $\Delta e_\varpi < 0.5$, then use the trained neural network as the first ϖ Weak classifiers CL_ϖ Otherwise, the neural network will be trained every iteration on the original basis N_u And check the weighted error rate of e, until the weighted error rate meets $\Delta e_\varpi < 0.5$ before stopping iteration. Finally, the weighted error rate Δe_ϖ As the AdaBoost framework ϖ Iterations CL_ϖ Error rate.

Step 5: Calculation u_ϖ As CL_ϖ Weight of.

$$u_\varpi = \frac{\ln\left(\frac{1-\Delta e_\varpi}{\Delta e_\varpi}\right)}{2} \tag{21}$$

Step 6: Update each sample to $w_{i,\varpi+1}$ The weight of is calculated according to the formula, normalized, and then executed $\varpi = \varpi + 1$, and skip to step 5 for the next iteration.

Step 7: BP weak classifier is established. judge $\varpi <$ Maximum Iterations max ϖ Whether it is true or not. If it is not true, the iteration is completed and the strong classifier has been established. Exit the algorithm. Otherwise, skip to Step 4.

For the established Adaboost-BP strong classifier, the joint output of each output node is:

$$Y_i = \arg\max \sum_{\varpi}^{\max \varpi} \log \frac{1}{u_\varpi} (CL_\varpi.out_\varpi = \varsigma) \tag{22}$$

among ς If it is 0 or 1, that is, the weight sum of the output is 1 and the weight sum of the output is 0. Select the maximum weight and the corresponding output as the output of the strong classifier.

After completing the training of Adaboost-BP classifier, it can be used for test sample recognition.

3 Method Test

3.1 Datasets

At present, the research on the combination of motion recognition technology and dance has just started, and the available dance data set is still relatively small. The open motion capture data set of Carnegie Mellon University, but the data set contains very little dance data, which cannot be specifically used for dance motion recognition research; The Dance DB dance data set published by the Virtual Reality Laboratory of the University of Cyprus can meet the requirements of dance action recognition research. Therefore, two dance data sets were used in the experiment, namely, the Dance DB data set and the folk dance data set produced by my laboratory. In the Dance DB dataset, each dance category uses emotion markers; The Folk Dance dance data set is divided into four groups in total. Each group contains a number of subdivisions of dance actions. The action categories are relatively rich, and each group of dance actions is relatively complex and challenging.

(1) DanceDB

There are 48 dance videos in the DanceDB dance dataset currently published by the Virtual Reality Laboratory of the University of Cyprus. The background and camera perspective in each dance video are fixed. The frame rate of the image is 20fps, and the size of each frame is 480 * 360. Although the data set currently contains a relatively small number of categories, there are challenges such as easily mixing moving objects and backgrounds in the video. It is an excellent dance action data set published in the field of dance action analysis research. Therefore, it can be used to measure the effectiveness of the algorithm proposed in this paper. There are 12 kinds of dance actions in the DanceDB dance data set, each of which is marked with an emotion tag as the category of this kind of dance action. The dance action categories in this dataset are: Afraid, Angry, Annoyed, Bored, Excited, Happy, Miserable, Pleased, Relaxed, Sad, Satisfied, Tired.

(2) FolkDance

The FolkDance dance data set is a dance data set produced by the laboratory itself. It uses the motion capture equipment Vicon to collect professional dance action videos. During the production of the entire data set, four groups of folk dance actions are designed according to the data set production plan and the final scheme discussed with dance experts. In view of the fact that the research on dance action recognition is still in its infancy, the production of the FolkDance dataset currently only considers the situation of single dance, and does not consider the changing stage background, props and other factors. In the specific process of dance video capture, we invited several dance majors to perform dance according to the grouping settings, while using the Vicon device to collect dance video data. A total of 84 dance videos were recorded, and the background and camera angle of view in each video were set to be fixed. The image in the video is uniformly set to a frame rate of 20 prints per second, and the size of each frame is 480 * 3600. This data set contains many types of dance actions, and the dance actions are complex, which is challenging for dance action recognition. Therefore, this data set can be used to verify the effectiveness of the dance action recognition algorithm proposed in this paper. The FolkDance dance data set mainly includes four groups of dances, namely, step double flower combination, lining flower combination, towel flower combination and flower combination.

3.2 Test Method

In order to verify the feasibility of the dance action recognition algorithm in this paper, we used cross validation to evaluate the algorithm in the experiment. Cross validation is a statistical method of cutting data samples into subsets. Its idea is to divide the original data set into training set and test set. Usually, the training set is used to train the classifier. After the training is completed, the test set is used to test the model obtained through training, and to evaluate the performance of the classifier, that is, the feasibility of the algorithm. K-fold cross validation is a common method of cross validation.

K fold cross validation: divide the data set into K groups, select one of them as the test set, and the remaining K-1 groups as the training set. Repeat the cross validation for K times, and select one group from them as the test set each time. Finally, the recognition accuracy of K times is taken as the final recognition result through the average cross validation. Generally, the value of K in the experiment is 10. On DanceDB, one person's dance data set is selected as the test set each time, and the rest three people's dance data set is selected as the training set. Repeat four times, and finally take the average result of the four times as the final result; For the FolkDance data set, the data of one person is also selected as the test set each time, and the data of the other two persons is used as the training set. Repeat three times, and finally average the final results of the three times.

3.3 Edge Contour Features

Use the research in chapter 1.2 to extract image edge contour features. Taking one of the video image samples as an example, the image edge contour features are shown in Fig. 3 below.

(a) Original drawing   (b) Edge contour feature drawing

Fig. 3. Edge contour feature extraction results

The outline area of the figure is 24.63 cm2; The contour perimeter is 32.27 cm;Hu moment similarity is 0.745.

3.4 Method Test Results

Combining all recognition results, calculate the action recognition accuracy of the two dance video data sets, and then calculate the average value. The results are shown in Table 1 below.

Table 1. Accuracy rate of action recognition

Method	Data set	Recognition accuracy/%	Average recognition accuracy/%
A Recognition Method Based on Edge Features	DanceDB	93.65	94.715
	FolkDance	95.78	
Identification Method Based on Improved Dense Trajectory	DanceDB	88.66	85.22
	FolkDance	81.78	
Identification method based on dual flow network	DanceDB	85.97	84.375
	FolkDance	82.78	
Recognition method based on bone joint point data	DanceDB	78.62	82.795
	FolkDance	86.97	

According to Table 1, the recognition accuracy of the article's method is relatively high, with an average of 94.715%. However, the average recognition accuracy of the improved dense trajectory based recognition method is 85.22%, the average recognition accuracy of the dual flow network based recognition method is 84.375%, and the average recognition accuracy of the bone joint number based recognition method is 82.795%. This indicates that the edge feature based recognition method proposed in this article has high recognition accuracy and can effectively achieve specific action recognition.

4 Conclusion

The recognition and analysis of dance movements has a very broad application prospect, and can play an important role in such aspects as dance video understanding, dance distance teaching and cultural protection. Nowadays, more and more advanced motion recognition algorithms and computing devices also help to achieve all of this. However, the research in this area is still very rare. This paper studies dance motion recognition based on edge features, and mainly completes the following work:

(1) Aiming at the research and analysis of the characteristics of dance movements at the same time, this paper proposes an effective method to extract edge features, analyzes the general steps of edge detection, and then takes Canny operator as an example, mainly introduces the steps of edge detection using Canny operator and the specific operations of each step Hu moment similarity characterizes the appearance and contour features of dance movements in the video.

(2) The AdaBoost framework algorithm is used to improve the BP neural network algo-rithm, and the AdaBoost enhanced BP neural network algorithm is proposed, which overcomes the problems that the traditional BP neural network is easy to fall into the local minimum and the convergence speed is slow.

(3) Another major contribution of this paper is the collection and production of dance data sets. For dance action recognition research, dance data sets play a very critical role in the research. We have specially produced a folk dance dataset. During the production process, we developed a detailed data set recording scheme, and discussed with professional dance experts about the production of dance data sets. In the later specific recording process, we used the Vicon motion capture system to invite different dance majors to record dance videos according to the dance group motion design. At the same time, considering that the research on dance action recognition is still in its infancy, and the dance action is too complex, our data set is recorded in a fixed scene and with a single person performing dance. At present, we have completed a total of three person times, four groups, and 84 dance action videos, as well as other single person time and multi category data sets for other dance research.

Although the method in this paper has finally achieved good action recognition results, there are still many areas worthy of improvement. In view of the complexity of dance movements, the dance data set we have produced at this stage only considers the situation of single person dance, and does not consider factors such as changing stage scenes. In the future, we will focus on more challenging research on dance action recognition such as scene changes and multi person dance, extract dance action features more suitable for complex backgrounds to better represent dance actions, and then make research results on dance action recognition that are more in line with actual needs such as real dance choreography.

Aknowledgement. Characteristics and Innovation of Grassroots Party Building in Xiangtan under the Background of Rural Revitalization (Project No. 2023C54).

References

1. Wang, F., Hu, R., Jin, Y.: Research on gesture image recognition method based on transfer learning. Procedia Comput. Sci. **187**(10), 140–145 (2021)
2. Jin, S.: Image recognition method for fault service action of tennis based on feature matching[J]. International Journal of Biometrics **13**(2/3), 150 (2021)
3. Sun, K., Zhang, B., Chen, Y., et al.: The facial expression recognition method based on image fusion and CNN. Integr. Ferroelectr. **217**(1), 198–213 (2021)
4. Yang, X., Liu, D., Liu, J., et al.: Follower: A Novel Self-Deployable Action Recognition Framework[J]. Sensors **21**(3), 950 (2021)
5. Toldinas, J., Venkauskas, A., Damaeviius, R., et al.: A novel approach for network intrusion detection using multistage deep learning image recognition. Electronics **10**(15), 1854 (2021)
6. Daradkeh, Y.I., Tvoroshenko, I., Gorokhovatskyi, V., et al.: Development of effective methods for structural image recognition using the principles of data granulation and apparatus of fuzzy logic. IEEE Access **9**(99), 13417–13428 (2021)
7. Chen, M., Wang, X., Luo, H., et al.: Learning to focus: cascaded feature matching network for few-shot image recognition. Sci. Chin. Inf. Sci. **64**(9), 192105 (2021)

8. Xiong, J., Yu, D., Liu, S., et al.: A review of plant phenotypic image recognition technology based on deep learning. Electronics **10**(1), 81 (2021)

9. Jin, L., Liang, H., Yang, C.: Sonar image recognition of underwater target based on convolutional neural network. Xibei Gongye Daxue Xuebao/J. Northwest. Polytechnical Univ. **39**(2), 285–291 (2021)

10. Tian, L., Xu, H., Zheng, X.: Research on fingerprint image recognition based on convolution neural network. Int. J. Biometrics **13**(1), 64–79 (2021)

11. Lyu, Z., Yu, Y., Samali, B., et al.: Back-propagation neural network optimized by K-fold cross-validation for prediction of torsional strength of reinforced concrete beam. Materials **15**(4), 1477 (2022)

12. Reza Kashyzadeh, K., Amiri, N., Ghorbani, S., et al.: Prediction of concrete compressive strength using a back-propagation neural network optimized by a genetic algorithm and response surface analysis considering the appearance of aggregates and curing conditions. Buildings **12**(4), 438 (2022)

13. Cong, Y.L., Hou, L.T., Wu, Y.C., et al.: Energy consumption prediction and diagnosis of heating ventilation and air conditioning system based on bidirectional LSTM method. In: 2022 International Conference on Computer Engineering and Artificial Intelligence (ICCEAI), pp. 633–636. IEEE (2022)

14. Jindal, H., Yadav, A., Sehgal, A., et al.: Geospatial landslide prediction–analysis & prediction from 2018-2022. J. Pharmaceutical Negative Results 2589–2599 (2023)

Author Index

Printed in the United States
by Baker & Taylor Publisher Services